T0134559

Topic Detection and Classification in Social Networks

Dimitrios Milioris

Topic Detection and Classification in Social Networks

The Twitter Case

Dimitrios Milioris
Massachusetts Institute of Technology
Cambridge, MA, USA

ISBN 978-3-319-88238-3   ISBN 978-3-319-66414-9   (eBook)
DOI 10.1007/978-3-319-66414-9

Printed on acid-free paper

This Springer imprint is published by Springer Nature
The registered company is Springer International Publishing AG
The registered company address is: Gewerbestrasse 11, 6330 Cham, Switzerland

The language given to me was Greek
My house poor on Homer's shores
My only care, my language on Homer's
shores . . .

The Axion Esti – Odysseas Elytis
Nobel Prize Winner, Poetry 1979

Preface

This book provides a novel method for topic detection and classification in social networks. The book addresses several research and technical challenges which are currently investigated by the research community, from the analysis of relations and communications between members of a community; quality, authority, relevance, and timeliness of the content; traffic prediction based on media consumption; and spam detection to security, privacy, and protection of personal information. Furthermore, the book discusses state-of-the-art techniques to address those challenges and provides novel techniques based on information theory and combinatorics, which are applied on real data obtained from Twitter. More specifically, the book:

- Detects topics from large text documents and extracts the main opinion without any human intervention
- Provides a language-agnostic method, which is not based on a specific grammar or semantics, e.g. machine learning techniques
- Compares state-of-the-art techniques and provides a smooth transition from theory to practice with multiple experiments and results

This book discusses dynamic networks, either social or delay-tolerant networks, and gives insight of specific methods for extracting prominent information along with methodology useful to students. It goes from theory to practice with experiments on real data.

Cambridge, MA, USA   Dimitrios Milioris
March 2017

Acknowledgments

It is a pleasure to thank the many people who made this book possible.

I would like to express my gratitude to my PhD supervisor, Dr. Philippe Jacquet, who was abundantly helpful and offered invaluable assistance, support, and guidance.

Deepest gratitude is also due to Prof. Wojciech Szpankowski, without whose knowledge and assistance this study would not have been successful.

I wish to offer my regards and blessings to all of my best friends in undergraduate and graduate level for helping me get through the difficult times in Athens, Crete, Paris, New York, and Boston and for all the emotional support, camaraderie, entertainment, and caring they provided: Stamatis Z., Marios P., Dimitris K., Dimitris A., George T., Christos T., Kostas C., Kostas P., Dani K., Jessica S., Alaa A., Emanuele M., Ioanna C., Vagelis V., Gérard B., Alonso S. Deepest gratitude goes to Stella, my star, who shows me the way.

I would like also to convey my thanks to the University of Crete, École Polytechnique, Inria, Bell Laboratories, Columbia University, and the Massachusetts Institute of Technology for providing the financial means and laboratory facilities throughout my career.

Lastly, and most importantly, I wish to thank my sister, Maria, and my parents, Milioris Spyridonas and Anastasiou Magdalini. They bore me, raised me, supported me, taught me, and loved me. To them I dedicate this book.

Cambridge, MA, USA   Dimitrios Milioris

Contents

Acronyms

AES	Advanced Encryption Standard
API	Application Program Interface
ASCII	American Standard Code for Information Interchange
BCS	Bayesian Compressive Sensing
BCS-GSM	Bayesian Compressive Sensing—Gaussian Scale Mixture
CS	Compressive Sensing
CT	Central Tweet
CVX	Convex Programming Solutions
DES	Data Encryption Standard
df-idft	Document Frequency-Inverse Document Frequency
DNA	Deoxyribonucleic Acid
DP	Document Pivot
DPurl	Document Pivot with URL
DynMC	Dynamic Matrix Completion
FP	Frequent Pattern
FPC	Fixed Point and Bregman Iterative Method
FPM	Frequent Pattern Matching
JC	Joint Complexity
JC+CS	Joint Complexity and Compressive Sensing
JCurl	Joint Complexity with URL
JCurl+CS	Joint Complexity with URL and CS
JCurl+PM	Joint Complexity with URL and PM
L1EQ-PD	ℓ_1-Norm Minimization Using the Primal-Dual Interior Point Method
LASSO	Least Absolute Shrinkage and Selection Operator
LDA	Latent Dirichlet Allocation
LSH	Locality-Sensitive Hashing
MC	Matrix Completion
OMP	Orthogonal Matching Pursuit
PAT tree	Patricia Tree
RIP	Restricted Isometry Property
SCAN	Structural Clustering Algorithm for Networks

SFPM	Soft Frequent Pattern Mining
SNSs	Social Networking Sites
ST	Suffix Tree
StOMP	Stagewise Orthogonal Matching Pursuit
SVT	Singular Value Thresholding
tf-idf	Term Frequency-Inverse Document Frequency
Triple DES	Triple Data Encryption Standard
URL	Uniform Resource Locator
VOIP	Voice over Internet Protocol
VPN	Virtual Private Network
WWW	World Wide Web

Chapter 1
Introduction

Abstract Social networks have undergone a dramatic growth in recent years. Such networks provide an extremely suitable space to instantly share multimedia information between individuals and their neighbours in the social graph. Social networks provide a powerful reflection of the structure, the dynamics of the society and the interaction of the Internet generation with both people and technology. Indeed, the dramatic growth of social multimedia and user generated content is revolutionizing all phases of the content value chain including production, processing, distribution and consumption. It also originated and brought to the multimedia sector a new underestimated and now critical aspect of science and technology, which is social interaction and networking. The importance of this new rapidly evolving research field is clearly evidenced by the many associated emerging technologies and applications, including (a) online content sharing services and communities, (b) multimedia communication over the Internet, (c) social multimedia search, (d) interactive services and entertainment, (e) health care and (f) security applications. It has generated a new research area called social multimedia computing, in which well-established computing and multimedia networking technologies are brought together with emerging social media research.

1.1 Dynamic Social Networks

Social networks have undergone a dramatic growth in recent years. Such networks provide an extremely suitable space to instantly share multimedia information between individuals and their neighbours in the social graph. Social networks provide a powerful reflection of the structure, the dynamics of the society and the interaction of the Internet generation with both people and technology. Indeed, the dramatic growth of social multimedia and user generated content is revolutionizing all phases of the content value chain including production, processing, distribution and consumption. It also originated and brought to the multimedia sector a new underestimated and now critical aspect of science and technology, which is social interaction and networking. The importance of this new rapidly evolving research field is clearly evidenced by the many associated emerging technologies and applications, including (a) online content sharing services and communities, (b) multimedia communication over the Internet, (c) social multimedia search, (d)

© Springer International Publishing AG 2018   1
D. Milioris, *Topic Detection and Classification in Social Networks*,
DOI 10.1007/978-3-319-66414-9_1

interactive services and entertainment, (e) health care and (f) security applications. It has generated a new research area called social multimedia computing, in which well established computing and multimedia networking technologies are brought together with emerging social media research.

Social networking services are changing the way we communicate with others, entertain and actually live. Social networking is one of the primary reasons why more people have become avid Internet users, people who until the emergence of social networks could not find interests in the Web. This is a very robust indicator of what is really happening online. Nowadays, users both produce and consume significant quantities of multimedia content. Moreover, their behaviour through online communities is forming a new Internet era where multimedia content sharing through Social Networking Sites (SNSs) is an everyday practice. More than 200 SNSs of worldwide impact are known today and this number is growing quickly. Many of the existing top web sites are either SNSs or offer some social networking capabilities.

Except for the major social networks with hundreds of millions of users that span in the entire world, there are also many smaller SNSs which are equally as popular as the major social networks within the more limited geographical scope of their membership, e.g. within a city or a country. There are also many vertically oriented communities that gather users around a specific topic and have many dedicated members on all continents.

Facebook is ranked among the most visited sites in the world, with more than 1.78 billion subscribed users to date. Moreover, Friendster is popular in Asia, Orkut in Brazil and Vkon-takte in Russia. On top of that, there are dozens of other social networks with vibrant communities, such as Vznet, Xing, Badoo, Netlog, Tuenti, Barrabes, Hyves, Nasza Klasa, LunarStorm, Zoo, Sapo, Daily-Motion and so on. There are also many vertically oriented communities which gather users around a specific topic, such as music, books, etc. LinkedIn with over 450 million users or Viadeo with 65 million users and Xing with 14 million users are mostly oriented in establishing professional connections between their users and initiate potential business collaborations.

The rapid growth in popularity of social networks has enabled large numbers of users to communicate, create and share content, give and receive recommendations, and, at the same time, it opened new challenging problems. The unbounded growth of content and users pushes the Internet technologies to its limits and demands for new solutions. Such challenges are present in all SNSs to a greater or lesser extent. Considerable amount of effort has already been devoted worldwide for problems such as content management in large scale collections, context awareness, multimedia search and retrieval, social graph modelling analysis and mining, etc.

1.1.1 The Twitter Social Network

Twitter is an online social networking service that enables users to send and read short messages of up to 140 characters called "tweets". Registered users can read and post tweets, but unregistered users can only read them. Users access Twitter through the website interface, SMS, or through a mobile device application. Twitter is one of the most popular social networks and micro-blogging service in the world, and according to its website it has more than 340 million active users connected by 24 billion links. In Twitter, "following" someone means that a user will have in his personal timeline other people's tweets (Twitter updates). "Followers" are people who receive other people's Twitter updates. Approximately 99.89% of the Twitter accounts have less than 3500 followers and followings. There are approximately 40 million accounts with less than 10 followers and followings, that is between 6% and 7% of all Twitter accounts. It is a social trend to ask followed accounts to follow back in Twitter. There is a limit at 2000 followings that starts growing after 1800 followers, which is the number of followings set by Twitter to prevent users monitoring too many accounts whereas they have no active role in Twitter. Approximately 40% of accounts have no followers and 25% have no followings. Twitter is interesting to be studied because it allows the information spread between people, groups and advertisers, and since the relation between its users is unidirectional, the information propagation within the network is similar to the way that the information propagates in real life.

1.2 Research and Technical Challenges

This section lists the main research challenges in social networks, which are currently being investigated by the research community.

- The analysis of relations and communications between members of a community can reveal the most influential users from a social point of view.
- As social networks will continue to evolve, the discovery of communities, users' interests [32], and the construction of specific social graphs from large scale social networks will continue to be a dynamic research challenge [85]. Research in dynamics and trends in social networks may provide valuable tools for information extraction that may be used for epidemic predictions or recommender systems [52, 87, 97].
- The information extracted from social networks proved to be a useful tool towards security. One example of an application related to security is the terrorism analysis, e.g. the analysis of the 9–11 terrorist network [106]. This study was done by gathering public information from major newspapers on the Internet and analyzed it by means of social networks [102]. Therefore, cyber surveillance for critical infrastructure protection is another major research challenge on social network analysis.

- Searching in blogs, tweets and other social media is still an open issue since posts are very small in size but numerous, with little contextual information. Moreover, different users have different needs when it comes to the consumption of social media. Real time search has to balance between quality, authority, relevance and timeliness of the content [105].
- Crowdsourcing systems gave promising solutions to problems that were unsolved for years. The research community nowadays is working by leveraging human intelligence to solve critical problems [57, 86], since social networks contain immense knowledge through their users. However, it is not trivial to extract that knowledge [50].
- Traffic prediction based on media consumption may be correlated between groups of users. This information can be used to dimension media servers and network resources to avoid congestion and improve the quality of experience and service.

   Content sharing and distribution needs will continue to increase. Mobile phones, digital cameras and other pervasive devices produce huge amounts of data which users want to distribute if possible in real time [15].
- Since users population and data production increase, spam and advertisements will continue growing [58]. In addition, the importance of social networks to influence the opinions of the users has to be protected with a mechanism that promotes trustworthy opinions that are relevant to businesses.
- As in every human community, online social communities face also critical social and ethical issues that need special care and delicate handling. Protection of personal information and many other problems need special attention [26].

   In order to address these challenges, we need to extract the relevant information from online social media in real time.

1.3 Problem Statement and Objectives

Topic detection and trend sensing is the problem of automatically detecting topics from large text documents and extracting the main opinion without any human intervention. This problem is of great practical importance given the massive volume of documents available online in news feeds, electronic mail, digital libraries and social networks.

Text classification is the task of assigning predefined labels or categories to texts or documents. It can provide conceptual views of document collections and has important applications in real-world problems. Nowadays, the documents which can be found online are typically organized by categories according to their subject, e.g. topics. Some widespread applications of topic detection and text classification are community detection, traffic prediction, dimensioning media consumption, privacy and spam filtering, as mentioned in Sect. 1.2.

By performing topic detection on social network communities, we can regroup users in teams and find the most influential ones, which can be used to build specific and strategic plans. Public information in social networks can be extracted by topic detection and classification and used for cyber surveillance in an automatic way in order to avoid the overload. Extracting an opinion from social networks is difficult, because users are writing in a way which does not have correct syntax or grammar and contains many abbreviations. Therefore, mining opinions in social networks can benefit by an automatic topic detection on really short and tremendous posts. By grouping users and adding labels to discussions or communities we are able to find their interests and tag people that share information very often. This information can be used to dimension media servers and network resources to avoid congestion and improve the quality of experience and service. Finally, by performing topic classification we can find similarities between posts of users that spread irrelevant information into the network and enable a spam filter to defend against that.

In this book we present a novel method to perform topic detection, classification and trend sensing in short texts. The importance of the proposed method comes from the fact that up to now, the main methods used for text classification are based on keywords detection and machine learning techniques. By using keywords or bag-of-words in tweets will often fail because of the wrongly or distorted usage of the words—which also needs lists of keywords for every language to be built— or because of implicit references to previous texts or messages. In general, machine learning techniques are heavy and complex and therefore are not good candidates for real-time text classification, especially in the case of Twitter where we have natural language and thousands of tweets per second to process. Furthermore machine learning processes have to be manually initiated by tuning parameters, and it is one of the main drawbacks for that kind of application. Some other methods are using information extracted by visiting the specific URLs on the text, which makes them a heavy procedure, since one may have limited or no access to the information, e.g. because of access rights, or data size. In this book we are trying to address the discussed challenges and problems of other state-of-the-art methods and propose a method which is not based on keywords, language, grammar or dictionaries, in order to perform topic detection, classification and trend sensing.

Instead of relying on words as most other existing methods which use bag-of-words or n-gram techniques, we introduce Joint Complexity (JC), which is defined as the cardinality of a set of all distinct common factors, subsequences of characters, of two given strings. Each short sequence of text is decomposed in linear time into a memory efficient structure called Suffix Tree (ST) and by overlapping two trees, in linear or sublinear average time, we obtain the JC defined as the cardinality of factors that are common in both trees. The method has been extensively tested for text generation by Markov sources of finite order for a finite alphabet. The Markovian generation of text gives a good approximation for natural text generation and is a good candidate for language discrimination. One key take-away from this approach is that JC is language-agnostic since we can detect similarities between two texts without being based on grammar and vocabulary. Therefore there is no need to build any specific dictionary or stemming process. JC can also be used to capture a change in topic within a conversation, as well as a change in the style of a specific writer of a text.

On the other hand, the inherent sparsity of the data space motivated us in a natural fashion the use of the recently introduced theory of Compressive Sensing (CS) [12, 18] driven by the problem of target localization [21]. More specifically, the problem of estimating the unknown class of a message is reduced to a problem of recovering a sparse position-indicator vector, with all of its components being zero except for the component corresponding to the unknown class where the message is placed. CS states that signals which are sparse or compressible in a suitable transform basis can be recovered from a highly reduced number of incoherent random projections, in contrast to the traditional methods dominated by the well-established Nyquist-Shannon sampling theory. The method works in conjunction with a Kalman filter to update the states of a dynamical system as a refinement step.

In this book we exploit datasets collected by using the Twitter streaming API, getting tweets in various languages and we obtain very promising results when comparing to state-of-the-art methods.

1.4 Scope and Plan of the Book

In this book, a novel method for topic detection, classification and trend sensing in Dynamic Social Networks is proposed and implemented. Such system is able to address the research and technical challenges mentioned in Sect. 1.2. The structure of this book is organized as follows.

First, Chap. 2 overviews the state-of-the-art of topic detection, classification and trend sensing techniques for online social networks. First, it describes the *document-pivot* and *feature-pivot* methods, along with a brief overview of the pre-processing stage of these techniques. Six state of the art methods: LDA, Doc-p, GFeat-p, FPM, SFPM, BNgram are described in detail, as they serve as the performance benchmarks to the proposed system.

In Chap. 3, we introduce the Joint Sequence Complexity method. This chapter describes the mathematical concept of the complexity of a sequence, which is defined as the number of distinct subsequences of the given sequence. The analysis of a sequence in subcomponents is done by suffix trees, which is a simple, fast, and low complexity method to store and recall subsequences from the memory. We define and use Joint Complexity for evaluating the similarity between sequences generated by different sources. Markov models well describe the generation of natural text, and their performance can be predicted via linear algebra, combinatorics and asymptotic analysis. We exploit Markov sources trained on different natural language datasets, for short and long sequences, and perform automated online sequence analysis on information streams in Twitter.

Then, Chap. 4 introduces the Compressive Sensing based classification method. Driven by the methodology of indoor localization, the algorithm converts the classification problem into a signal recovery problem, so that CS theory can be applied. First we employ Joint Complexity to perform topic detection and build signal vectors. Then we apply the theory of CS to perform topic classification by recovering an indicator vector, while reducing significantly the amount of information from tweets. Kalman filter is introduced as a refinement step for the update of the process, and perform users and topics tracking.

Chapter 5 presents the extension of Joint Complexity and Compressive Sensing on three additional research subjects that have been studied during this book, (a) classification encryption via compressed permuted measurement matrices, (b) dynamic classification completeness based on Matrix Completion and (c) cryptography based on the Eulerian circuits of original texts.

Finally, Chap. 6 presents the concluding remarks and gives directions for future work and perspectives.

Chapter 2
Background and Related Work

Abstract Topic detection and tracking aims at extracting topics from a stream of textual information sources, or documents, and to quantify their "trend" in real time. These techniques apply on pieces of texts, i.e. posts, produced within social media platforms. Topic detection can produce two types of complementary outputs: cluster output or term output are selected and then clustered. In the first method, referred to as *document-pivot*, a topic is represented by a cluster of documents, whereas in the latter, commonly referred to as *feature-pivot*, a cluster of terms is produced instead. In the following, we review several popular approaches that fall in either of the two categories. Six state-of-the-art methods: Latent Dirichlet Allocation (LDA), Document-Pivot Topic Detection (Doc-p), Graph-Based Feature-Pivot Topic Detection (GFeat-p), Frequent Pattern Mining (FPM), Soft Frequent Pattern Mining (SFPM), BNgram are described in detail, as they serve as the performance benchmarks to the proposed system.

2.1 Introduction

Topic detection and tracking aims at extracting topics from a stream of textual information sources, or documents, and to quantify their "trend" in real time [3]. These techniques apply on pieces of texts, i.e. posts, produced within social media platforms. Topic detection can produce two types of complementary outputs: cluster output or term output are selected and then clustered. In the first method, referred to as *document-pivot*, a topic is represented by a cluster of documents, whereas in the latter, commonly referred to as *feature-pivot*, a cluster of terms is produced instead. In the following, we review several popular approaches that fall in either of the two categories. Six state-of-the-art methods: Latent Dirichlet Allocation (LDA), Document-Pivot Topic Detection (Doc-p), Graph-Based Feature-Pivot Topic Detection (GFeat-p), Frequent Pattern Mining (FPM), Soft Frequent Pattern Mining (SFPM), BNgram are described in detail, as they serve as the performance benchmarks to the proposed system.

© Springer International Publishing AG 2018   9
D. Milioris, *Topic Detection and Classification in Social Networks*,
DOI 10.1007/978-3-319-66414-9_2

2.2   Document-Pivot Methods

Simple *document-pivot* approaches cluster documents by leveraging some similarity metric between them. A recent work [83] follows this direction to provide a method for "breaking news" detection in Twitter. Tweets retrieved using targeted queries or hashtags are converted into a *bag-of-words* representation weighted with boosted *tf-idf* (term frequency-inverse document frequency) emphasizing important entities such as names of countries or public figures. *Bag-of-words* of a text is its representation as the set of its words, disregarding grammar and even word order but keeping multiplicity. Tweets are then incrementally merged in clusters by considering the textual similarity between incoming tweets and existing clusters. Similar approaches based on textual similarity and *tf-idf* can be found in literature [6, 78]. Among them, the method discussed in [6] classifies tweets as referring to real-world events or not. The classifier is trained on a vast variety of features including social aspects (e.g. number of mentions) and other Twitter-specific features. An important drawback of the method is the need for manual annotation of training and test samples.

Dimensions other than text can also be used to improve the quality of clustering. TwitterStand [89] uses a "leader–follower" clustering algorithm that takes into account both textual similarity and temporal proximity. Each cluster center is represented using a centroid *tf-idf* vector and the average post-time of the tweet in the cluster. A similarity metric based on both elements and on the number of shared hashtags allows incremental merging of new tweets with existing clusters. The main disadvantages of this method are the sensitivity to noise (which is a known problem for *document-pivot* methods [25]) and fragmentation of clusters. It needs a manual selection of trusted information providers and periodic defragmentation runs to mitigate such effects. The goal in a large corpus is to detect the first document discussing a given topic like in [82]. A new story is created by a document having low similarity with all previously detected clusters. Locality sensitive hashing is used for fast retrieval of nearest neighbours for the incoming document.

In conclusion, document-pivot methods have two main problems, which are: (a) segmentation of classification groups, and (b) depend on arbitrary thresholds for the classification of an incoming tweet.

2.3   Feature-Pivot Methods

Feature-pivot methods are closely related to topic models in natural language processing, namely statistical models used to build sets of terms which are representative of the topics occurring in a corpus of documents. Most of the state-of-the-art static topic models are based on the Latent Dirichlet allocation (LDA) [8], which is described in Sect. 2.4.3. Even though LDA extensions for dynamic data have been proposed [7], alternative approaches trying to capture topics through the detection of term burstiness have been studied [91], mainly in the context of news media mining. The idea behind those methods is that "breaking news", unlike other discussion topics, happen to reach a fast peak of attention from routine users as soon as they are tweeted or posted [53, 108]. Accordingly, the common framework which underlies most of the approaches in this category first identifies bursty terms and then clusters them together to produce topic definitions.

The diffusion of the services over social media and detection of bursty events had been studied in generic document sets. The method presented in [25], for instance, detects bursty terms by looking their frequency in a given time window. Once the bursty terms are found, they are clustered using a probabilistic model of cooccurrence. The need for such a global topic term distribution restricts this approach to a batch mode of computation. Similar methods were tested for topic detection in social media, such as in the Twitter, but with additional emphasis on the enrichment of the obtained topics with non-bursty but relevant terms, URLs and locations [59].

Graph-based approaches detect term clusters based on their pairwise similarities. The algorithm proposed in [90] builds a term cooccurrence graph, whose nodes are clustered using a community detection algorithm based on betweenness centrality, which is an indicator of a node's centrality in a network and is equal to the number of shortest paths from all vertices to all others that pass through that node. Additionally, the topic description is enriched with the documents which are most relevant to the identified terms. Graphs of short phrases, rather than of single terms, connected by edges representing similarity have also been used [54]. Graph-based approaches have also been used in the context of collaborative tagging systems with the goal of discovering groups of tags pertaining to topics of social interest [80].

Signal processing approaches have also been explored in [103], which compute *df-idf* (a variant of *tf-idf*) for each term in each considered time slot, and then apply wavelet analysis on consecutive blocks. The difference between the normalized entropy of consecutive blocks is used to construct the final signal. Relevant terms which are bursty are extracted by computing the autocorrelation of the signal and heuristically learning and determining a threshold to detect new bursty terms. Also in this case, a graph between selected terms is built based on their cross-correlation and it is then clustered to obtain event definitions. The Discrete Fourier Transform is used in [31], where the signal for each term is classified according to its power and periodicity. Depending on the identified class, the distribution of appearance of a term in time is modelled using Gaussian distributions. The Kullback–Leibler divergence (as a relative entropy) between the distributions is then used to determine clusters and increase the computational complexity of the method.

The knowledge of the community leads to even more sophisticated approaches. In a recent work [13] a PageRank-like measure is used to identify important users on the Twitter social network. Such centrality score is combined with a measure of term frequency to obtain a measure for each term. Then, clustering on a correlation graph of bursty terms delineates topic boundaries.

These methods are based on the analysis of similarities between terms and often give wrong correlation of topics, with their main disadvantage being the use of dictionaries and stemming processes.

2.4   Related Work

In general the topic detection methods use preprocessing techniques. Next, we define all the components of the topic detection process. In Sect. 2.4.1, we present the problem statement and define some basic terminology. Then in Sect. 2.4.2, we describe the data preprocessing and in the following sections we present six methods that take in as input the preprocessed data and output the detected topics.

2.4.1   Problem Definition

We address the task of detecting topics in real-time from social media streams. To keep our approach general, we consider that the stream is made of short pieces of text generated by social media users, e.g. posts, messages or tweets in the specific case of Twitter. Posts are formed by a sequence of words or terms, and each one is marked with the timestamp of creation. A plethora of methods have a user-centred scenario in which the user starts up the detection system by providing a set of seed terms that are used as initial filter to narrow down the analysis only to the posts containing at least one of the seed terms. Additionally, there exists an assumption that the time frame of interest (can be indefinitely long) and a desired update rate are provided (e.g. detect new trending topics every 15 min). The expected output of the algorithm is a topic, defined as a headline and a list of terms, delivered at the end of each time slot determined by the update rate. This setup fits well many real-world scenarios in which an expert of some domain has to monitor specific topics or events being discussed in social media [17, 87]. For instance, this is the case for computational journalism in which the media inquirer is supposed to have enough knowledge of the domain of interest to provide initial terms to perform an initial filtering of the data stream. Even if it requires an initial human input, this framework still remains generic and suitable to any type of topic or event.

2.4.2   Data Preprocessing

The content of user generated messages could be unpredictably noisy. In many works, in order to reduce the amount of noise before the proper topic detection is executed, the raw data extracted through the seed terms filter is subjected to three preprocessing steps.

- Tokenization: In a raw post, terms can be combined with any sort of punctuation and hyphenation and can contain abbreviations, typos or conventional word variations. The Twokenizer tool [78] is used to extract bags of cleaner terms from the original messages by removing stopwords and punctuation, compressing redundant character repetitions, and removing mentions, i.e., IDs or names of other users included in the text for messaging purposes.

- Stemming: In information retrieval, stemming is the process of reducing inflected words to their root (or stem), so that related words map to the same stem. This process naturally reduces the number of words associated to each document, thus simplifying the feature space. Most techniques use an implementation of the Porter stemming algorithm [84].
- Aggregation: Topic detection methods based on word or n-grams cooccurrences, or any other type of statistical inference, suffer in the absence of long documents. This is the case of social media, where user-generated content is typically in the form of short posts. In information retrieval it is a common practice to partially address this problem by concatenating different messages together to produce documents of larger size. Large documents construction is based on two strategies. The first strategy involves temporal aggregation that concatenates together N messages, whose generation date is contiguous. The second strategy involves a similarity-based aggregation which attaches to a message all the near-duplicate messages posted in the same time slot, identified through an efficient document clustering method [82], which is also used by one of the examined topic detection algorithms presented in Sect. 2.4.4.

Determining the effect of such preprocessing algorithms on the quality of the final topic detection is difficult to assess, and it has not been much investigated so far. For instance, the aggregation of posts in super-documents could, on the one hand, help to improve the word cooccurrence statistic but, on the other hand, introduces the risk of grouping terms related to different topics, and to reveal false cooccurrence.

2.4.3   Latent Dirichlet Allocation

Topic extraction in textual *corpus* can be addressed through probabilistic topic models. In general, a topic model is a Bayesian model which associates with each document a probability distribution over the topics, where each topic is in turn a probability distribution. The Latent Dirichlet Allocation (LDA) [8] is the best known and most widely used topic model. According to LDA, every document is considered as a bag of terms, which are the only observed variables in the model. The topic distribution per document and the term distribution per topic are instead hidden variable and have to be estimated through Bayesian inference. The Collapsed Variational Bayesian inference algorithm [95], which is an LDA variant, is computationally efficient, more accurate than standard variational Bayesian inference for LDA, and has given rise to many independent implementations already available in the literature. LDA requires the expected number of topics k as an input. The estimation of the optimal k, although possible through the use of non-parametric methods [94], falls beyond the scope of this book.

2.4.4 Document-Pivot Topic Detection

The second method discussed here, is an instance of a classical Topic Detection and Tracking method which uses a document-pivot approach (Doc-p). It works as follows:

- First, the method performs online clustering of posts. It computes the cosine similarity of the *tf-idf* [88] representation of an incoming post with all other posts processed so far. If the best cosine similarity is above some threshold θ_{tf-idf}, it assigns the item to the same cluster as its best match; otherwise, it creates a new cluster with the new post as its only item. The best matching tweet is efficiently retrieved by Locality Sensitive Hashing (LSH).
- Then, it filters out clusters with item count smaller than θ_n.
- For each cluster c, it computes a score as follows:

$$score(c) = \sum_{doc \in C} \sum_{word \in doc} \exp(-p(word))$$

  The probability of appearance of a single term $p(word)$ is estimated from a reference dataset that has been collected from Twitter, mentioned in Sect. 2.4.5. Thus, less frequent terms contribute more to the score of the cluster.
- Clusters are sorted according to their score and the top clusters are returned. LSH can rapidly provide the nearest neighbours with respect to cosine similarity in a large collection of documents. An alternative to LSH is to use inverted indices on the terms which appear in the tweets and then compute the cosine similarity between the incoming document and the set of documents that have a significant term overlap with it; however, the use of LSH is much more efficient as it can provide the nearest neighbours with respect to cosine similarity directly.

In practice, when posts are as short as tweets, the similarity of two items is usually either close to zero or close to one (between 0.8 and 1.0). This observation makes setting θ_{tf-idf} relatively easy to 0.5. Due to this phenomenon, items grouped together by this procedure are usually, but not always, near-duplicates (e.g. ReTweets). Therefore, it is clear that topics produced by this method will be fragmented, i.e. the same topic may be represented by different sets of near duplicate tweets. To begin dealing with this issue, we present methods for aggregating as described in Sect. 2.4.2.

2.4.5 Graph-Based Feature-Pivot Topic Detection

The Graph-Based Feature-Pivot Topic Detection method (GFeat-p) is a first of a series of *feature-pivot* methods, where the clustering step is performed via the Structural Clustering Algorithm for Networks (SCAN) [107], which is in general applied to network analysis. Apart from detecting communities of nodes, SCAN provides a list of hubs (vertices that bridge many clusters), each of which may be connected to a set of communities. For SCAN applied to topic detection, the nodes in the network correspond to terms and the communities correspond to topics. The detected hubs are terms which are related to more than one topic, and effectively provides an explicit link between topics. That would not be possible to achieve with a common partition clustering algorithm. The terms to be clustered, is a subset of terms present in the *corpus*, applied to independent reference corpus selected with randomly collected tweets [78]. For each of the terms in the reference corpus, the likelihood of appearance $p(w|corpus)$ is estimated as follows:

$$p(w|corpus) = \frac{N_w(corpus) + \delta}{N(corpus)}$$

where $N_w(corpus)$ is the number of appearances of term w in the corpus, N(corpus) is the number of terms in corpus (including repetition) and δ is a constant (typically set to 0.5) that is included to regularize the probability estimate (i.e. to ensure that a new term that does not appear in the corpus is not assigned a probability of 0). The most important terms in the new corpus are determined by computing the ratio of the likelihoods of appearance in the two corpora for each term, as follows:

$$\frac{p(w|corpus_{new})}{p(w|corpus_{ref})}$$

The terms with the highest ratio are expected to be related to the most actively discussed topics in the *corpus*. Once the high ratio terms are identified, a term graph is constructed and the SCAN graph-based clustering algorithm is applied to extract groups of terms, each of which is considered to be a distinct topic. More specifically, the algorithm steps are the following:

- Selection: The top K terms are selected using the ratio of likelihoods measure and will be used as the nodes for a graph G.
- Linking: The nodes of G are connected using a term linking strategy. According to a preselected similarity measure for pairs of terms all pairwise similarities are computed.

    Moreover, each term is linked with its k nearest neighbours (kNN approach) or all pairs of nodes that have similarity higher than ϵ (ϵ-based approach) are linked.
- Clustering: The SCAN algorithm is applied to the graph; a topic is generated for each of the detected communities. A hub may be linked to more than one topic or community.

The term linking step is crucial for the success of the method. Unfortunately, there is no straightforward method for determining a priori the best similarity measure or node linking strategy to be used. Additionally, it can be expected that the graph construction tuning parameters will need to vary for datasets with different topic granularities and levels of internal topic connectivity.

2.4.6   Frequent Pattern Mining

As it was mentioned in the previous section, a problem with the *feature-pivot* methods is that terms clustering relies on pairwise similarities which are based on cooccurrences number only. In the case of closely interconnected topics which share a relatively large number of terms, this procedure most likely produces general topics. An option to deal with this issue is to take into account the simultaneous cooccurrence between more than two terms. This motivation leads naturally to consider the use of frequent set of items, a well-defined technique in transaction mining for topic detection to determine which items are likely to cooccur in a set of transactions [27].

In the social media context, an item is any term *w* mentioned in a post (excluding stop words, punctuation tokens, etc.). The transaction is the post, and the transaction set are all posts that occur in a time slot *T*. The number of times that any given set of terms occurs in the time slot is defined as its support, and any set of items that meets a minimum support is called a pattern. The initial challenge is to apply highly scalable Frequent Pattern (FP) detection to each time slot to identify the most frequent terms or patterns. These terms are used to characterize the topics that best illustrate the underlying social interactions. Below, we describe these two processing steps, FP detection and ranking.

1. FP detection: First, the algorithm constructs a term list according to their frequency [30, 56]. Then, for each time slot an FP-Tree sorts the patterns according to their cooccurrences and their support. Finally, the FP-tree structures are aggregated and analysed to produce association rules on the transaction set.
2. FP ranking: Once a set of frequent patterns has been extracted from the dataset, they are ranked and the top *N* results are returned as candidate topics. The challenge is to rank patterns such that terms in the candidate topics are sufficiently related and with enough diversity to cover the different underlying subjects of conversation in the social interactions. A common way to rank patterns is to simply use the support of a given pattern; the more often a set of terms cooccurs, the more likely it can be considered relevant as a topic.

2.4.7   Soft Frequent Pattern Mining

In Sect. 2.4.6 a frequent pattern mining approach for topic detection was described. It provided an elegant solution to the problem of *feature-pivot* methods that takes into account only pairwise cooccurrences between terms in the case of corpus with densely interconnected topics. Section 2.4.5 examined only pairwise cooccurrences, where frequent pattern mining examines cooccurrences between any number of terms, typically larger than two. A question that naturally arises is whether it is possible to formulate a method that lies between these two extremes. Such a method would examine cooccurrence patterns between sets of terms with cardinality larger than two, like frequent pattern mining does, but it would be less strict by not requiring that all terms in these sets cooccur frequently. Instead, in order to ensure topic cohesiveness, it would require that large subsets of the terms grouped together, but not necessarily all, cooccur frequently, resulting in a "soft" version of frequent pattern mining.

The proposed approach (SFPM) works by maintaining a set of terms S, on which new terms are added in a greedy manner, according to how often they cooccur with the terms in S. In order to quantify the cooccurrence match between a set S and a candidate term t, a vector D_S for S and a vector D_t for the term t are maintained, both with dimension n, where n is the number of documents in the collection. The i-th element of D_S denotes how many of the terms in S cooccur in the i-th document, whereas the i-th element of D_t is a binary indicator that represents if the term t occurs in the i-th document or not. Thus, the vector D_t for a term t that frequently cooccurs with the terms in set S will have a high cosine similarity to the corresponding vector D_S. Note that some of the elements of D_S may have the value $|S|$, meaning that all items in S occur in the corresponding documents, whereas other may have a smaller value indicating that only a subset of the terms in S cooccur in the corresponding documents. For a term that is examined for expansion of S, it is clear that there will be some contribution to the similarity score also from the documents in which not all terms cooccur, albeit somewhat smaller compared to that documents in which all terms cooccur. The "soft" matching between a term that is considered for expansion and a set S is achieved. Finding the best matching term can be done either using exhaustive search or some approximate nearest neighbour scheme such as LSH. As mentioned, a greedy approach that expands the set S with the best matching term is used, thus a criterion is needed to terminate the expansion process. The termination criterion clearly has to deal with the cohesiveness of the generated topics, meaning that if not properly set, the resulting topics may either end up having too few terms or really being a mixture of topics (many terms related to possibly irrelevant topics). To deal with this, the cosine similarity threshold $\theta(S)$ between S and the next best matching term is used. If the similarity is above the threshold, the term is added, otherwise the expansion process stops. This threshold is the only parameter of the proposed algorithm and is set to be a function of the cardinality of S. In particular a sigmoid function of the following form is used:

$$\theta(S) = 1 - \frac{1}{1 + \exp((|S| - b)/c)}$$

The parameters b and c can be used to control the size of the term clusters and how soft the cooccurrence constraints will be. Practically, the values of b and c are set so that the addition of terms when the cardinality of S is small is easier (the threshold is low), but addition of terms when the cardinality is larger is harder. A low threshold for the small values of $|S|$ is required so that it is possible for terms that are associated to different topics and therefore occur in more documents rather than to ones corresponding to the non-zero elements of D_S to join the set S. The high threshold for the larger values of $|S|$ is required so that S does not

grow without limit. Since a set of topics is required—rather than a single topic—
the greedy search procedure is applied as many times as the number of considered
terms, each time initializing S with a candidate term. This will produce as many
topics as the set of terms considered, many of which will be duplicates, thus a
post-process of the results is needed to remove these duplicates. To limit the search
procedure in reasonable limits the top n terms with the highest likelihood-ratio are
selected by following the methodology in (2.4.5).

When the "soft" frequent pattern matching algorithm runs for some time, the
vector D_S may include too many non-zero entries filled with small values, especially
if some very frequently occurring term has been added to the set. This may have the
effect that a term may be deemed relevant to S because it cooccurs frequently only
with a very small number of terms in the set rather than with most of them. In order
to deal with this issue, after each expansion step, any entries of D_S that have a value
smaller than $|S|/2$ are reset to zero. The most relevant documents for a topic can be
directly read from its vector D_S: the ones with the highest document counts.

2.4.8   BNgram

Both the frequent itemset mining and soft frequent itemset mining approaches
attempted to take into account the simultaneous cooccurences between more than
two terms. However, it is also possible to achieve a similar result in a simpler way
by using n-grams. This naturally groups together terms that cooccur and it may be
considered to offer a first level of term grouping. Using n-grams makes particularly
sense for Twitter, since a large number of the status updates in Twitter are just
copies or retweets of previous messages, so important n-grams will tend to become
frequent.

Additionally, a new feature selection method is introduced. The changing
frequency of terms over time as a useful source of information to detect emerging
topics is taken into account. The main goal of this approach is to find emerging
topics in post streams by comparing the term frequencies from the current time slot
with those of preceding time slots. The $df - idf_t$ metric which introduces time to the
classic *tf-idf* score is proposed. Historical data to penalize those topics which began
in the past and are still popular in the present, and which therefore do not define new
topics have been used.

The term indices, implemented using Lucene, are organized into different time
slots. In addition to single terms, the index also considers bigrams and trigrams.
Once the index is created, the $df - idf_t$ score is computed for each n-gram of the
current time slot i based on its document frequency for this time slot and penalized
by the logarithm of the average of its document frequencies in the previous t time
slots:

$$score_{df-idf_t} = \frac{df_i + 1}{\log\left(\frac{\sum_{j=1}^{t} df_{i-j}}{t} + 1\right) + 1}$$

In addition, a boost factor is considered to raise the importance of proper nouns (persons, locations and organizations, in our case) using a standard named entity recognizer [22], as they are essential terms in most discussed stories. The use of this factor is similar to [83], where the authors highlight the importance of such words for grouping results. The selected values for this factor are based on the best values from the experiments of the previous work, being boost = 1.5 in case the n-gram contains a named entity and boost = 1 otherwise. As a result of this process, a ranking of n-grams is created sorted by their df-idft scores. A single n-gram is often not very informative, but a group of them often offers interesting details of a story. Therefore, a clustering algorithm to group the most representative n-grams into clusters, each representing a single topic is used. The clustering is based on distances between n-grams or clusters of n-grams. From the set of distances, those not exceeding a distance threshold are assumed to represent the same topic. The similarity between two n-grams is defined as the fraction of posts that contain both of them. Every n-gram is initially assigned to its own singleton cluster, then following a standard "group average" hierarchical clustering algorithm [72] to iteratively find and merge the closest pair of clusters. When an n-gram cluster is joined to another, the similarities of the new cluster to the other clusters are computed as the average of the similarities of the combined clusters. The clustering is repeated until the similarity between the nearest un-merged clusters falls below a fixed threshold θ, producing the final set of topic clusters for the corresponding timeslot.

The main contribution of this approach is the use of the temporal dimension of data to detect emerging stories. There are other similar approaches on term weighting considering the temporal dimension of data but most of them suffer from several shortcomings. For instance, in [91] two methods of finding "peaky topics" are presented. They find the peak terms for a time-slot compared to the rest of the corpus, whereas each slot is compared to the immediately previous time slots. If some topic is discussed at several different times, their approach could miss this since the defining words would be highly frequent in the whole corpus. In addition, their approach only uses single words which often seem to be too limited to identify stories. Finally, their use of the whole corpus is less suitable for real-time analysis.

2.5   Chapter Summary

This chapter overviewed the state of the art of topic detection, classification and trend sensing techniques for online social networks. First, it described the *document-pivot* and *feature-pivot* methods, along with a brief overview of the pre-processing stage of these techniques. Six state-of-the-art methods: LDA, Doc-p, GFeat-p, FPM, SFPM, BNgram were described in detail, as they serve as the performance benchmarks to the proposed system.

Chapter 3
Joint Sequence Complexity: Introduction and Theory

Abstract In this chapter we study joint sequence complexity and we introduce its applications for topic detection and text classification, in particular source discrimination. The mathematical concept of the complexity of a sequence is defined as the number of distinct factors of it. The Joint Complexity is thus the number of distinct common factors of two sequences. Sequences containing many common parts have a higher Joint Complexity. The extraction of the factors of a sequence is done by suffix trees, which is a simple and fast (low complexity) method to store and retrieve them from the memory. Joint Complexity is used for evaluating the similarity between sequences generated by different sources and we will predict its performance over Markov sources. Markov models describe well the generation of natural text, and their performance can be predicted via linear algebra, combinatorics and asymptotic analysis. This analysis follows in this chapter. We exploit datasets from different natural languages, for both short and long sequences, with promising results on complexity and accuracy. We performed automated online sequence analysis on information streams in Twitter.

3.1 Introduction

In this chapter we study joint sequence complexity and we introduce its applications for topic detection and text classification, in particular source discrimination. The mathematical concept of the complexity of a sequence is defined as the number of distinct factors of it. The Joint Complexity is thus the number of distinct common factors of two sequences. Sequences containing many common parts have a higher Joint Complexity. The extraction of the factors of a sequence is done by suffix trees, which is a simple and fast (low complexity) method to store and retrieve them from the memory. Joint Complexity is used for evaluating the similarity between sequences generated by different sources and we will predict its performance over Markov sources. Markov models describe well the generation of natural text, and their performance can be predicted via linear algebra, combinatorics and asymptotic analysis. This analysis follows in this chapter. We exploit datasets from different natural languages, for both short and long sequences, with promising results on complexity and accuracy. We performed automated online sequence analysis on information streams in Twitter.

© Springer International Publishing AG 2018   21
D. Milioris, *Topic Detection and Classification in Social Networks*,
DOI 10.1007/978-3-319-66414-9_3

3.2 Sequence Complexity

In the last decades, several attempts have been made to capture mathematically the concept of the "complexity" of a sequence. In [48], the sequence complexity was defined as the number of different factors contained in a sequence. If X is a sequence and $I(X)$ its set of factors (distinct substrings), then the cardinality $|I(X)|$ is the complexity of the sequence. For example, if $X = aabaa$, then $I(X) = \{v, a, b, aa, ab, ba, aab, aba, baa, aaba, abaa, aabaa\}$, and $|I(X)| = 12$, where v denotes the empty string. Sometimes the complexity of the sequence is called the I-Complexity (IC) [5]. The notion is connected with quite deep mathematical properties, including the rather elusive concept of randomness in a string [34, 55, 75].

3.3 Joint Complexity

In general, the information contained in a string cannot be measured in absolute and a reference string is required. The concept of Joint Complexity (JC) has been introduced in [36], as a metric of similarity between two strings. JC method is the number of different factors, which are common in two sequences. In other words the JC of sequence X and Y is equal to $J(X, Y) = |I(X) \cap I(Y)|$. We denote $J_{n,m}$ the average value of $J(X, Y)$ when X and Y have length n and m, respectively. In this work we study its growth when the lengths are the same, $n = m$.

JC method is an efficient way of evaluating the degree of similarity between two sequences. For example, the genome sequences of two dogs will contain more common factors than the genome sequences of a dog and a cat. Similarly, two texts written in the same language have more common factors than texts written in very different languages. JC method is also greater when languages have similarities (e.g. French and Italian) than when they differ significantly (e.g. French and English). Furthermore, texts written in the same language but on different topics (e.g. law and cooking) have smaller JC than texts on the same topic (e.g. medicine). Therefore, JC method is a pertinent tool for automated monitoring of social networks. This requires a precise analysis of the JC method, discussed in this work as well as in [43], together with some experimental results, confirming usefulness of the joint string complexity for short text discrimination.

In [36] it is proved that the JC of two texts of length n built from two different binary memoryless sources grows like

$$\gamma \frac{n^\kappa}{\sqrt{\alpha \log n}}$$

for some $\kappa < 1$ and $\gamma, \alpha > 0$ which depend on the parameters of the two sources. When the sources are identical, i.e. when their parameters are identical, but the text still being independently generated, then the JC growth is $O(n)$, hence $\kappa = 1$. When the texts are identical (i.e. $X = Y$), then the JC is identical to the I-Complexity and it grows as $\frac{n^2}{2}$ [48]. Therefore JC method can already be used to detect "copy–paste" parts between the texts. Indeed the presence of a common factor of length $O(n)$ would inflate the JC by a term $O(n^2)$.

We should point out that experiments demonstrate that for memoryless sources the JC estimate

$$\gamma \frac{n^k}{\sqrt{\alpha \log n}}$$

converges very slowly. Therefore, JC is not really meaningful even when $n \approx 10^9$. In this work we derive second order asymptotics for JC of the following form

$$\gamma \frac{n^k}{\sqrt{\alpha \log n + \beta}}$$

for some $\beta > 0$. Indeed it turns out that for text where $n < 100$ and $\log n < \beta$, this new estimate converges more quickly than the estimate

$$\gamma \frac{n^k}{\sqrt{\alpha \log n}}$$

thus it can be used for short texts, like tweets. In fact, our analysis indicates that JC can be refined via a factor for

$$P\left(\frac{1}{\alpha \log n + \beta} \right)$$

appearing in the JC, where P is a specific polynomial determined via saddle point expansion. This additional term further improves the convergence for small values of n, and also same periodic factors of small amplitude appear when the source parameters satisfy some specific and very unfrequent conditions.

In this work we extend the JC estimate to Markov sources of any order on a finite alphabet. Although Markov models are no more realistic than memoryless sources, say, for a DNA sequence, they seem to be fairly realistic for text generation [43]. An example of Markov simulated text for different order is shown in Table 3.1 from "The Picture of Dorian Gray".

In view of these facts, we can use the JC to discriminate between two identical/non-identical Markov sources [109]. We introduce the discriminant function as follows

$$d(X, Y) = 1 - \frac{1}{\log n} \log J(X, Y)$$

Table 3.1 Markov simulated text from the book "The Picture of Dorian Gray"

Markov order	Text simulated by the given Markov order
3	"Oh, I do yourse trought lips whose-red from to his, now far taked. If Dorian, Had kept has it, realize of him. Ther chariten suddenial tering us. I don't belige had keption the want you are ters. I am in the when mights horry for own that words is Eton of sould the Oh, of him to oblige mere an was not goods"
4	"Oh, I want your lives. It is that is words the right it his find it man at they see merely fresh impulses. But when you have you, Mr. Gray, a fresh impulse of mine. His sad stifling round of a regret. She is quite devoted forgot an arrowed"
5	"Oh, I am so sorry. When I am in Lady Agatha's black book that the air of God, which had never open you must go. I am painting, and poisons us. We are such a fresh impulse of joy that he has done with a funny looked at himself unspotted"

for two sequences X and Y of length n. This discriminant allows us to determine whether or not X and Y are generated by the same Markov source by verifying whether

$$d(X, Y) = O(1/\log n) \to 0 \ \text{ or } \ d(X, Y) = 1 - \kappa + O(\log \log n / \log n) > 0$$

respectively when the length of X and Y are both equal to n. In this work we concentrate mainly on the analysis of the JC method, however, we also present some experimental evidence of how useful our discriminant is for real texts.

In Fig. 3.1, we compared the JC of a real English text with simulated texts of the same length written in French, Greek, Polish and Finnish (all language is transcribed in the Latin alphabet, simulated from a Markov source of order 3). It is easy to see that even for texts smaller in length than a thousand words, one can discriminate between these languages. By discriminating, we mean that the JC between texts of different languages drops significantly in comparison to JC for texts of the same language. The figure shows that Polish, Greek and Finnish are further from English than French is. On the other hand, in Fig. 3.2, we plot the similarity between real and simulated texts in French, Greek, Polish, English and Finnish.

In Polish, the second part of the text shifts to a different topic, and we can see that the method can capture this difference. Clearly, the JC of such texts grows like $O(n)$ as predicted by theory. In fact, computations show that with Markov models of order 3 for English versus French we have $\kappa = 0.44$; versus Greek: $\kappa = 0.26$; versus Finnish: $\kappa = 0.04$; and versus Polish: $\kappa = 0.01$, which is consistent with the results in Fig. 3.1, except for the low value of κ where the convergence to the asymptotics regime is slower. In fact, they agree with the actual resolution of Eq. (3.7), which contains the transition to an asymptotics regime. A comparison between different topics or subjects is presented in Fig. 3.3. We test four texts from books on constitutional and copyright law as well as texts extracted from two cookbooks. As we can see, the method can well distinguish the differences, and shows increased similarity for the same topic.

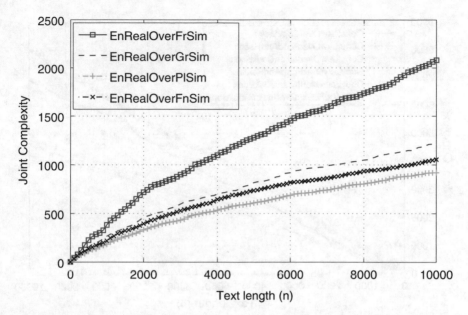

Fig. 3.1 Joint Complexity of real English text versus simulated texts in French, Greek, Polish and Finnish

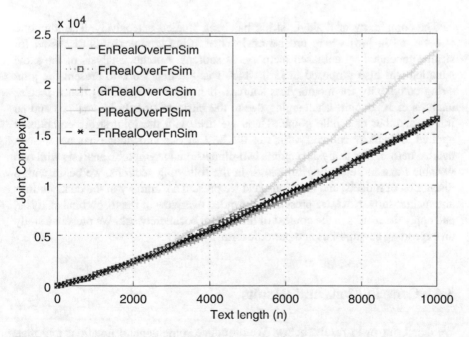

Fig. 3.2 Joint complexity of real and simulated texts (3rd Markov order) in the English, French, Greek, Polish and Finnish languages

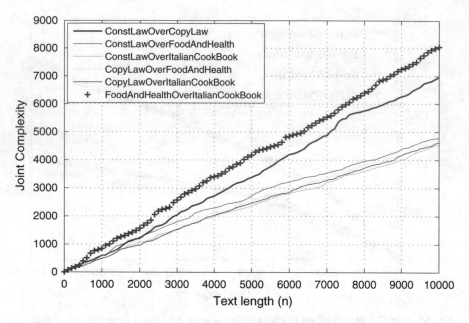

Fig. 3.3 Joint complexity of real text from a variety of books spanning constitutional and copyright law to healthy cooking and recipes from the cuisine of Italy

The complexity of a single string has been studied extensively. The literature is reviewed in [48] where precise analysis of string complexity is discussed for strings generated by unbiased memoryless sources. Another analysis of the same situation was also proposed in [36] which was the first work to present the joint string complexity for memoryless sources. It is evident from [36] that a precise analysis of JC is quite challenging due to the intricate singularity analysis and an infinite number of saddle points. Here, we deal with the joint string complexity applied to Markov sources, which to the best of our knowledge has never been tackled before. The analysis requires two-dimensional asymptotic analysis with two variable Poisson generating functions. In the following sections, we begin with a discussion of models, and notations, and we present a summary of the contributions and main results. Next, we present an extended overview of the theoretical analysis, and apply the results in the context of the Twitter social network. We present a study on expending asymptotics and periodic terms.

3.4 Contributions and Results

We detail our main results below. We introduce some general notations and then present a summary.

3.4.1 Models and Notations

Let ω and σ be two strings over a finite alphabet $\mathscr{A}$ (e.g. $\mathscr{A} = \{a, b\}$). We denote by $|\omega|_\sigma$ the number of times σ occurs as a factor in ω (e.g. $|abbba|_{bb} = 2$). By convention $|\omega|_v = |\omega| + 1$, where v is the empty string, because the empty string is prefix and suffix of ω and also appears between the characters, and $|\omega|$ is the length of the string.

Throughout we denote by X a string (text) and we plan to study its complexity. We also assume that its length $|X|$ is equal to n. Then the string complexity is $I(X) = \{\omega : |X|_\omega \geq 1\}$. Observe that

$$|I(X)| = \sum_{\sigma \in \mathscr{A}^*} 1_{|X|_\sigma \geq 1},$$

where 1_A is the indicator function of a Boolean A. Notice that $|I(X)|$ is equal to the number of nodes in the associated suffix tree of X [40, 48, 92]. We will come back on the suffix tree in Sect. 3.8.

Let X and Y be two sequences (not necessarily of the same length). We have defined the *Joint Complexity* as the cardinality of the set $J(X, Y) = I(X) \cap I(Y)$. We have

$$|J(X, Y)| = \sum_{\sigma \in \mathscr{A}^*} 1_{|X|_\sigma \geq 1} \times 1_{|Y|_\sigma \geq 1}.$$

We now assume that the strings X and Y are respectively generated by two *independent Markov sources* of order r, so-called source 1 and source 2. We will only deal here with Markov of order 1, but extension to arbitrary order is straightforward. We assume that source i, for $i \in \{1, 2\}$ has the transition probabilities $P_i(a|b)$ from term b to term a, where $(a, b) \in \mathscr{A}^r$. We denote by $\mathbf{P}_1$ (resp. $\mathbf{P}_2$) the transition matrix of Markov source 1 (resp. source 2). The stationary distributions are respectively denoted by $\pi_1(a)$ and $\pi_2(a)$ for $a \in \mathscr{A}^r$.

Let X_n and Y_m be two strings of respective lengths n and m, generated by Markov source 1 and Markov source 2, respectively. We write $J_{n,m} = \mathbf{E}(|J(X_n, Y_m)|) - 1$ for the joint complexity, i.e. omitting the empty string.

3.4.2 Summary of Contributions and Results

Definition: We say that a matrix $\mathbf{M} = [m_{ab}]_{(a,b)\in\mathscr{A}^2}$ is *rationally balanced* if $\forall (a, b, c) \in \mathscr{A}^3$: $m_{ab} + m_{ca} - m_{cb} \in \mathbb{Z}$, where $\mathbb{Z}$ is the set of integers. We say that a positive matrix $\mathbf{M} = [m_{ab}]$ is *logarithmically rationally balanced* when the matrix $\log^*(\mathbf{M}) = [\ell_{ab}]$ is rationally balanced, where $\ell_{ab} = \log(m_{ab})$ when $m_{ab} > 0$ and $\ell_{ab} = 0$ otherwise. Furthermore, we say that two matrices $\mathbf{M} = [m_{ab}]_{(a,b)\in\mathscr{A}^2}$ and $\mathbf{M}' = [m'_{ab}]$ are *logarithmically commensurable* when matrices $\log^*(\mathbf{M})$ and $\log^*(\mathbf{M}')$ are commensurable, that is, there exist a nonzero pair of reals (x, y) such that $x \log^*(\mathbf{M}) + y \log^*(\mathbf{M}')$ is rationally balanced.

We now present our main theoretical results in a series of theorems each treating different cases of Markov sources. Most of the mathematics were primary introduced in [43] and then developed in [46] and [66].

Theorem 3.1 *Joint Complexity over same sources: Consider the average joint complexity of two texts of length n generated by the same general stationary Markov source, that is, $\mathbf{P} := \mathbf{P}_1 = \mathbf{P}_2$.*

(i) *[Noncommensurable Case.] Assume that $\mathbf{P}$ is not logarithmically rationally balanced. Then*

$$J_{n,n} = \frac{2\log 2}{h}n + o(1) \tag{3.1}$$

where h is the entropy rate of the source.

(ii) *[Commensurable Case.] Assume that $\mathbf{P}$ is logarithmically rationally balanced. Then there is $\epsilon > 0$ such that:*

$$J_{n,n} = \frac{2\log 2}{h}(1 + Q_0(\log n)) + O(n^{-\epsilon})$$

where $Q_0(.)$ is a periodic function of small amplitude.

Now we consider different sources, i.e. $\mathbf{P}_1 \neq \mathbf{P}_2$ that are not the same and have respective transition matrices $\mathbf{P}_1$ and $\mathbf{P}_2$. The transition matrices are on $\mathscr{A}^r \times \mathscr{A}^r$. For a tuple of complex numbers (s_1, s_2) we write $\mathbf{P}(s_1, s_2)$ for a matrix whose (a, b)-th coefficient is $(\mathbf{P}_1(a|b))^{-s_1}(\mathbf{P}_2(a|b))^{-s_2}$, with the convention that if one of the $P_i(a|b) = 0$ then the (a, b) coefficient of $\mathbf{P}(s_1, s_2) = 0$.

We first consider the case when matrix $\mathbf{P}(s_1, s_2)$ is nilpotent [33], that is, for some K the matrix $\mathbf{P}^K(s_1, s_2) = 0$.

Theorem 3.2 *If $\mathbf{P}(s_1, s_2)$ is nilpotent, then there exists γ_0 and $\epsilon > 0$ such that $\lim_{n\to\infty} J_{n,n} = \gamma_0 := \langle \mathbf{1}(\mathbf{I} - \mathbf{P}(0, 0))^{-1}|\mathbf{1}\rangle$ where $\mathbf{1}$ is the unit vector and $\langle \cdot|\cdot\rangle$ is the inner product.*

This result is not surprising since the common factors are uniformly bounded in length, therefore form a finite set.

Indeed the common factors can only occur in a finite window of size K at the beginning of the strings and this will be developed in the proofs.

Throughout, now we assume that $\mathbf{P}(s_1, s_2)$ is not nilpotent. We denote by $\mathscr{K}$ the set of real tuple (s_1, s_2) such that $\mathbf{P}(s_1, s_2)$ has the main eigenvalue $\lambda(s_1, s_2)$ equal to 1. Let

$$\kappa = \min_{(s_1, s_2) \in \mathscr{K}} \{-s_1 - s_2\}$$

$$(c_1, c_2) = \arg\min_{(s_1, s_2) \in \mathscr{K}} \{-s_1 - s_2\}.$$

with $\kappa < 1$.

Lemma 3.1 *We have either $c_1 > 0$ or $c_2 > 0$ and $(c_1, c_2) \in [-1, 0]^2$.*

Theorem 3.3 *Assume* $P(s_1, s_2)$ *is not nilpotent and either* $c_1 > 0$ *or* $c_2 > 0$. *We only handle* $c_2 > 0$, *the case* $c_1 > 0$ *being obtained by symmetry.*

(i) *[Noncommensurable Case.] We assume that* P_2 *is not logarithmically balanced. Let* $c_0 < 0$ *such that* $(c_0, 0) \in \mathscr{K}$. *There exists* γ_1 *and* $\epsilon > 0$ *such that*

$$J_{n,n} = \gamma_1 n^{-c_0}(1 + O(n^{-\epsilon})) \tag{3.2}$$

(ii) *[Commensurable Case.] Let now* P_2 *be logarithmically rationally balanced. There exists a periodic function* $Q_1(.)$ *of small amplitude such that*

$$J_{n,n} = \gamma_1 n^{-c_0}(1 + Q_1(\log n) + O(n^{-\epsilon})).$$

The case when both c_1 and c_2 are between -1 and 0 is the most intricate to handle.

Theorem 3.4 *Assume that* c_1 *and* c_2 *are between* -1 *and* 0.

(i) *[Noncommensurable Case.] When* P_1 *and* P_2 *are not logarithmically commensurable matrices, then there exist* α_2, β_2 *and* γ_2 *such that*

$$J_{n,n} = \frac{\gamma_2 n^{\kappa}}{\sqrt{\alpha_2 \log n + \beta_2}} \left(1 + O\left(\frac{1}{\log n}\right)\right). \tag{3.3}$$

(ii) *[Commensurable Case.] Let* P_1 *and* P_2 *be logarithmically commensurable matrices. Then there exists a double periodic function* $Q_2(.)$ *of small amplitude such that:*

$$J_{n,n} = \frac{\gamma_2 n^{\kappa}}{\sqrt{\alpha_2 \log n + \beta_2}} \left(1 + Q_2(\log n) + O\left(\frac{1}{\log n}\right)\right).$$

3.5 Proofs of Contributions and Results

In this section we present proofs of our main results.

3.5.1 An Important Asymptotic Equivalence

We have the identity:

$$J_{n,m} = \sum_{w \in \mathscr{A}^* - \{v\}} P(w \in I(X_n)| \geq 1) \cdot P(w \in I(Y_n) \geq 1). \tag{3.4}$$

We know that from [20, 40] there is a closed formula for

$$\sum_n P(|X_n|_w \geq 1)z^n = \frac{P_1(w)z}{(1-z)D_w(z)}$$

which is, in the memoryless case,

$$D_w(z) = (1-z)(1 + A_w(z)) + P_1(w)z^{|w|}$$

where $P(w)$ is the probability that w is a prefix of X'_n and $A_w(z)$ is the "autocorrelation" polynomial of word w [40]. For the Markov source, we omit the expression which carries extra indices which track to the Markov correlations for the starting symbols of the words. A complete description of the parameters can be found in [20, 40].

Although it is a closed formula, this expression is not easy to manipulate. To make the analysis tractable we notice that $w \in I(X_n)$ is equivalent to the fact that w is at least a prefix of one of the n suffices of X_n.

If the suffices would have been n independent infinite strings, then $P(w \in I(X_n))$ would be equal to $1 - (1 - P_1(w))^n$ whose generating function is

$$\frac{P1(z)z}{(1-z)(1 - z + P_1(w)z)}$$

which would be the same as

$$\frac{P_1(w)z}{(1-z)(D_w(z))}$$

if we set in $D_w(z)$ with $A_w(z) = 0$ and identify $z^{|w|}$ to z.

We define $I_1(n)$ (resp. $I_2(n)$) as the set of prefixes of n independent strings built on source 1 (resp. 2). Let

$$C_{n,m} = \sum_{w \in \mathscr{A}^* - \{v\}} P(w \in I_1(n))P(w \in I_2(n))$$

As it is claimed in [20, 40] we prove that

Lemma 3.2 *There exists $\epsilon > 0$*

$$J_{n,n} = C_{n,n}(1 + O(n^{-\epsilon})) + O(1). \tag{3.5}$$

The proof is not developed in [20, 40] and seems to be rather complicated. It can be found in [44].

3.5.2 Functional Equations

Let $a \in \mathscr{A}$. We denote

$$C_{a,m,n} = \sum_{w \in a\mathscr{A}^*} P(w \in I_1(n)) P(w \in I_2(n))$$

where $w \in a\mathscr{A}^*$ for $a \in \mathscr{A}$ means that w starts with symbol a. Notice that $C_{a,m,n} = 0$ when $n = 0$ or $m = 0$. Using the Markov nature of the string generation, the quantity $C_{a,n,m}$ for $n, m \geq 1$ satisfies the following recurrence for all $a, b \in \mathscr{A}$

$$C_{b,n,m} = 1 + \sum_{a \in \mathscr{A}} \sum_{n_a, m_a} \binom{n}{n_a} \binom{m}{m_a}$$

$$\times (P_1(a|b))^{n_a} (1 - P_1(a|b))^{n - n_a}$$

$$\times (P_2(a|b))^{m_a} (1 - P_2(a|b))^{m - m_a} C_{a,n_a,m_a} \, ,$$

where n_a is an integer smaller or equal to n (resp. m_a) and denotes the number of strings among n (resp. m), independent strings from source 1 (resp. 2) which starts with symbol b, n strings starting with b and n_a tracks the number of such string that starts with ba. Quantity m_a is the counterpart for source 2. The *unconditional* average $C_{n,m}$ satisfies for $n, m \geq 2$

$$C_{n,m} = 1 + \sum_{a \in \mathscr{A}} \sum_{n_a, m_a} \binom{n}{n_a} \binom{m}{m_a} \pi_1^{n_a}(a)(1 - \pi_1(a))^{n - n_a}$$

$$\times \pi_2^{m_a}(a)(1 - \pi_2(a))^{m - m_a} C_{a,n,m}.$$

since $\pi_i(a)$ is the probability that a string from source i starts with symbol a.

We introduce the double Poisson transform of $C_{a,n,m}$ as

$$C_a(z_1, z_2) = \sum_{n,m \geq 0} C_{a,n,m} \frac{z_1^n z_2^m}{n! m!} e^{-z_1 - z_2} \tag{3.6}$$

which translates the recurrence (in the formula above n_a tracks the number) into the following functional equation:

$$C_b(z_1, z_2) = (1 - e^{-z_1})(1 - e^{-z_2})$$

$$+ \sum_{a \in \mathscr{A}} C_a (P_1(a|b)z_1, P_2(a|b)z_2). \tag{3.7}$$

Furthermore, the cumulative double Poisson transform

$$C(z_1, z_2) = \sum_{n,m \geq 0} C_{n,m} \frac{z_1^n z_2^m}{n! m!} e^{-z_1 - z_2} \tag{3.8}$$

satisfies

$$C(z_1, z_2) = (1 - e^{-z_1})(1 - e^{-z_2})$$

$$+ \sum_{a \in \mathscr{A}} C_a(\pi_1(a)z_1, \pi_2(a)z_2) . \tag{3.9}$$

3.5.3 Double DePoissonization

The asymptotics of the coefficient $C_{n,m}$ are extracted from the asymptotics of function $C(z_1, z_2)$ where $\Re(z_1, z_2) \to \infty$. This is an extension of DePoissonization theorems of [41, 43, 92], and are used to prove the following lemma.

Lemma 3.3 (DePoissonization) *When n and m tend to infinity:*

$$C_{n,m} = C(n, m)(1 + O(n^{-1}) + O(m^{-1})) .$$

This equivalence is obtained by proving some growth properties of $C(z_1, z_2)$ when (z_1, z_2) are complex numbers; such properties are stated in [43].

3.5.4 Same Markov Sources

We first present a general result when the Markov sources are identical: $\mathbf{P}_1 = \mathbf{P}_2 = \mathbf{P}$. In this case (3.7) can be rewritten with $c_a(z) = C_a(z, z)$:

$$c_b(z) = (1 - e^{-z})^2 + \sum_{a \in \mathscr{A}} c_a (P(a|b)z) . \tag{3.10}$$

This equation is directly solvable by the Mellin transform $c_a^*(s) = \int_0^\infty c_a(x)x^{s-1}dx$ defined for $-2 < \Re(s) < -1$. For all $b \in \mathscr{A}$ we find [92]

$$c_b^*(s) = (2^{-s} - 2)\Gamma(s) + \sum_{a \in \mathscr{A}} (P(a|b))^{-s} c_a^*(s) . \tag{3.11}$$

Introducing $c^*(s) = \int_0^\infty C(z, z)z^{s-1}dz$ [24], and the property of Mellin transform $\int f(ax)x^{s-1} = a^{-s} \int f(x)x^{s-1}dx$. The definition domain of $c^*(s)$ is $\Re(s) \in (-2, -1)$, and the Mellin transform $c^*(s)$ of $C(z, z)$ becomes

$$c^*(s) = (2^{-s} - 2)\Gamma(s) + \sum_{a \in \mathscr{A}} (\pi(a))^{-s} c_a^*(s) .$$

Thus

$$c^*(s) = (2^{-s} - 2)\Gamma(s) \left(1 + \langle \mathbf{1}(\mathbf{I} - \mathbf{P}(s))^{-1} | \boldsymbol{\pi}(s) \rangle \right) \tag{3.12}$$

where $\mathbf{1}$ is the vector of dimension $|\mathscr{A}|$ made of all 1's, $\mathbf{I}$ is the identity matrix, and $\mathbf{P}(s) = \mathbf{P}(s, 0) = \mathbf{P}(0, s)$, $\boldsymbol{\pi}(s)$ is the the vector made of coefficients $\pi(a)^{-s}$ and $\langle .|. \rangle$ denotes the inner product.

By applying the methodology of Flajolet [23, 92], the asymptotics of $c(z)$ for $|\arg(z)| < \theta$ is given by the residues of the function $c^*(s)z^{-s}$ occurring at $s = -1$ and $s = 0$. They are respectively equal to

$$\frac{2\log 2}{h} z \quad \text{and} \quad -1 - \langle \mathbf{1}(\mathbf{I} - \mathbf{P}(0, 0))^{-1}\boldsymbol{\pi}(0)\rangle.$$

The first residue comes from the singularity of $(\mathbf{I} - \mathbf{P}(s))^{-1}$ at $s = -1$. This led to the formula expressed in Theorem 3.1(i). When $\mathbf{P}$ is logarithmically rationally balanced then there are additional poles on a countable set of complex numbers s_k regularly spaced on the same imaginary axes containing -1 and such that $\mathbf{P}(s_k)$ has eigenvalue 1. These poles contribute to the periodic terms in Theorem 3.1(ii).

Computations on the trained transition matrix show that a Markov model of order 3 for English text has entropy of 0.944221, while French text has an entropy of 0.934681, Greek text has an entropy of 1.013384, Polish text has an entropy of 0.665113 and Finnish text has an entropy of 0.955442. This is consistent with Fig. 3.2.

3.5.5   Different Markov Sources

In this section we identify the constants in Theorems 3.3 and 3.4 with the assumption $\mathbf{P}_1 \neq \mathbf{P}_2$. We cannot obtain a functional equation for $C_a(z, z)$'s, and we thus have to deal with two variables z_1 and z_2. We define the double Mellin transform $C_a^*(s_1, s_2) = \int_0^\infty \int_0^\infty C_a(z_1, z_2)z_1^{s_1-1}z_2^{s_2-1}dz_1 dz_2$ and similarly the double Mellin transform $C^*(s_1, s_2)$ of $C(z_1, z_2)$. And thus we should have the identity

$$C_b^*(s_1, s_2) = \Gamma(s_1)\Gamma(s_2) \tag{3.13}$$
$$+ \sum_{a \in \mathscr{A}} (P_1(a|b))^{-s_1} (P_2(a|b))^{-s_2} C_a^*(s_1, s_2)$$

which leads to

$$C^*(s_1, s_2) = \Gamma(s_1)\Gamma(s_2) \left(1 + \langle \mathbf{1}(\mathbf{I} - \mathbf{P}(s_1, s_2))^{-1}|\boldsymbol{\pi}(s_1, s_2)\rangle\right) \tag{3.14}$$

where $\boldsymbol{\pi}(s_1, s_2)$ denotes the vector composed of coefficients $\pi_1(a)^{-s_1}\pi_2(a)^{-s_2}$. In fact to define the Mellin transform we need to apply it to

$$C(z_1, z_2) - \frac{\partial}{\partial z_1}C(0, z_2)z_1 e^{-z_1} - \frac{\partial}{\partial z_2}C(z_1, 0)z_2 e^{-z_2}$$

which leads to exponentially decaying but we omit this technical detail, which is
fully described in [43]. The original value $C(z_1, z_2)$ is obtained via the inverse Mellin
transform

$$C(z_1, z_2) = \frac{1}{(2i\pi)^2} \int \int C(s_1, s_2) z_1^{-s_1} z_2^{-s_2} ds_1 ds_2 \qquad (3.15)$$

thus

$$C(z, z) = \frac{1}{(2i\pi)^2} \int_{\Re(s_1)=\rho_1} \int_{\Re(s_2)=\rho_2} C^*(s_1, s_2) z^{-s_1-s_2} ds_1 ds_2 . \qquad (3.16)$$

where (ρ_1, ρ_2) belongs to the definition domain of $C^*(s_1, s_2)$; $\rho \in (-2, -1) \times (-2, -1)$.

We denote $L(s)$ be the function of complex s such that $\mathbf{P}(s, L(s))$ has eigenvalue
1 or equivalently such that $(\mathbf{I} - \mathbf{P}(s_1, s_2))^{-1}$ ceases to exist. The function $L(s)$ is
meromorphic and has several branches; one branch describes the set $\mathscr{K}$ when s is
real. Now to evaluate the double integral (3.16) we move the line of integration with
respect to s_2 from ρ_2 to some $M > 1$ while fixing the value of s_1 and collecting on
the way the residues on all the poles encountered. In particular, the dominant residue
at $s_2 = L(s)$ contributes

$$C(z, z) = \frac{1}{2i\pi} \int_{\Re(s_1)=\rho_1} \mu(s_1) \Gamma(s_1) \Gamma(L(s_1)) z^{-s_1-L(s_1)} ds_1$$

$$+ O(z^{\rho_1-M}) \qquad (3.17)$$

where $\mu(s)$ is the residue of $\langle \mathbf{1}(\mathbf{I} - \mathbf{P}(s, s_2))^{-1} \boldsymbol{\pi}(s_1, s_2) \rangle$ at point $(s, L(s))$, that is,

$$\mu(s_1) = \frac{1}{\frac{\partial}{\partial s_2} \lambda(s_1, s_2)} \langle \mathbf{1} | \boldsymbol{\zeta}(s_1, s_2) \rangle \langle \mathbf{u}(s_1, s_2) | \boldsymbol{\pi}(s_1, s_2) \rangle \big|_{s_2=L(s_1)} .$$

where $\lambda(s_1, s_2)$ is the eigenvalue which has value 1 at $(s, L(s))$ and $\mathbf{u}(s_1, s_2)$ and
$\boldsymbol{\zeta}(s_1, s_2)$ are respectively the left and right eigenvectors with the convention that
$\langle \boldsymbol{\zeta}(s_1, s_2) | \mathbf{u}(s_1, s_2) \rangle = 1$.

The expression is implicitly a sum since the function $L(s)$ is meromorphic, but
we retain only the branch where $\lambda(s_1, s_2)$ is the main eigenvalue of $\mathbf{P}(s_1, s_2)$ that
contributes to the leading term in the expansion of $C(z, z)$. For more details see [43]
where the analysis is specific to a case where one of the sources, namely source 2,
is memoryless uniform, i.e. $\mathbf{P}_2 = \frac{1}{|\mathscr{A}|} \mathbf{1} \otimes \mathbf{1}$.

The next step consists in moving the integration line for s_1 from ρ_1 to c_1 which
corresponds to the position where function $-s_1 - L(s_1)$ (actually equal to κ) attains
the minimum value. We only consider the case when $L(c_1) = c_2 < 0$ (the other
case is obtained by symmetry). The poles are due to the function $\Gamma(.)$. The first

pole encountered is $s_1 = -1$ but this pole cancels with the technical arrangement discussed earlier.

We do not work on the simple case, i.e. when $c_1 > 0$. We meet the second pole at $s = 0$ and the residue is equal to $\mu(0)\Gamma(c_0)z^{-c_0}$ since $L(0) = c_0$. This quantity turns out to be the leading term of $C(z, z)$ since the integration on $\Re(s_1) = c_1$ is $O(z^\kappa)$. This proves Theorem 3.3. When $\mathbf{P}_2$ is logarithmically balanced, there exists ω such that $\lambda(s, L(s) + ik\omega) = 1$ for $k \in \mathbb{Z}$ and the terms $z^{c_0 + ik\omega}$ lead to a periodic contribution.

The difficult case is when $-1 < c_1 < 0$. In this case, $C(z, z) = O(z^\kappa)$ but to find precise estimates one must use the saddle point method [23], at $s = c_1$ since the integration is of the form $\int_{\Re(s)} = c_1 f(s) \exp(-(s + L(s))A)ds$, where $f(s) = \mu(s)\Gamma(s)\Gamma(L(s))$, and $A = \log z \to \infty$. We naturally get an expansion when $\Re(z) \to \infty$

$$C(z, z) = \frac{e^{\kappa \log z} \mu(c_1)}{\sqrt{(\alpha_2 \log z + \beta_2)}} \left(1 + O\left(\frac{1}{\sqrt{\log z}}\right)\right)$$

with $\alpha_2 = L''(c_1)$ and $\beta_2 = \frac{\mu'(c_1)}{\mu(c_1)}$. In fact, the saddle point expansion is extendible to any order of $\frac{1}{\sqrt{\log z}}$. This proves Theorem 3.4 in the general case. However, in the case when $\mathbf{P}_1$ and $\mathbf{P}_2$ are logarithmically commensurable, the line $\Re(s_1) = c_1$ contains an infinite number of saddle points that contribute in a double periodic additional term.

Example: Assume that we have a binary alphabet $A = \{a, b\}$ with memory 1, and transition matrices $P_1 = \begin{bmatrix} 0.5 & 0.5 \\ 0.5 & 0.5 \end{bmatrix}$ and $P_2 = \begin{bmatrix} 0.2 & 0.8 \\ 0.8 & 0.2 \end{bmatrix}$

The numerical analysis gives

$$C_{n,n} = 13.06 \frac{n^{0.92}}{\sqrt{1.95 \log n + 56.33}} \implies C_{n,n} = 3.99(\log 2)n.$$

Inspired from the general results about the asymptotic digital trees and suffix tree parameters distribution we conjecture the following [40, 45, 74].

Conjecture 3.1

(i) The variance $V_{n,n}$ of the joint complexity of two random texts of same length n generated by two Markov sources is of order $O(n^\kappa)$ when $n \to \infty$.

(ii) The normalized distribution of the joint complexity $\mathbf{J}_{n,n}$ of these two texts tends to the normal distribution when $n \to \infty$.

Remark By "normalized distribution" we mean the distribution of $\frac{\mathbf{J}_{n,n} - J_{n,n}}{\sqrt{V_{n,n}}}$.

3.6 Expending Asymptotics and Periodic Terms

The estimate

$$J_{n,n} = \gamma \frac{n^k}{\sqrt{\alpha \log n + \beta}} \left(1 + Q(\log n) + O\left(\frac{1}{\log n}\right)\right)$$

which appears in the case of a different Markov source comes from a saddle point analysis. The potential periodic terms $Q(\log n)$ occur in a case where the Kernel $\mathscr{K}$ shows an infinite set of saddle points. It turns out that the amplitude of the periodic terms is of the order of

$$\Gamma\left(\frac{2i\pi}{\log |A|}\right)$$

i.e. of the order of 10^{-6} for binary alphabet, but it rises when $|A|$ increases. For example, when $|A| \geq 26$ such as in the Latin alphabet used in English (including spaces, commas and other punctuation) we get an order within 10^{-1}.

Figure 3.4 shows the number of common factors from two texts generated from two memoryless sources. One source is a uniform source over the 27 Latin symbols (such source is so-called monkey typing), the second source takes the statistic of letters occurrence in English. The trajectories are obtained by incrementing each text one by one. Although not quite significant, the logarithmic oscillations appear in the trajectories. We compare this with the expression

$$\gamma \frac{n^k}{\sqrt{\alpha \log n + \beta}}$$

without the oscillation terms which are actually

$$13.06 \frac{n^{0.92}}{\sqrt{1.95 \log n + 73.81}}.$$

In fact it turns out that the saddle point expression has a poor convergence term since the $O\left(\frac{1}{\log n}\right)$ is indeed in $\frac{1}{\alpha \log n + \beta}$ made poorer since the latter does not make less than $\frac{1}{\beta}$ for the text length range that we consider. But the saddle point approximation leads to the estimate factor $P_k((\alpha \log n + \beta)^{-1})$ of

$$J_{n,n} = \gamma \frac{n^k}{\sqrt{\alpha \log n + \beta}} (1 + P_k((\alpha \log n + \beta)^{-1})) + O\left(\frac{1}{(\log n)^{k+\frac{1}{2}}}\right) + Q(\log n)$$

$$(3.18)$$

where $P_k(x) = \sum_{j=1}^{k} A_j x^j$ is a specific series polynomial of degree k. The error term is thus in $(\alpha \log n + \beta)^{-k-\frac{1}{2}})$ but is not uniform for k. Indeed, the expansion

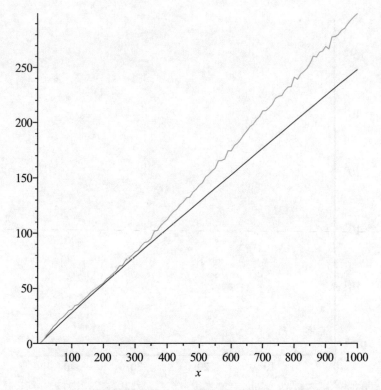

Fig. 3.4 Joint Complexity (*y* axis) of memoryless English text (*x* axis) versus monkey typing. The first order theoretical average is shown in *red* (*cont. line*)

polynomial P_k diverges when $k \to \infty$. Therefore for a given value of x there is an optimal value of k which minimizes the error term whose relative order is given by $|A_{k+1}x^{k+1} + A_{k+2}x^{k+2}|$, since the series of the A_i's have alternated signs.

Figure 3.5 shows the different error terms versus k for $x = \frac{1}{\beta}$. The optimal is reached for $k - 5$ with a relative error of 10^{-3}. Figure 3.6 shows the new theoretical average with $P_5((\alpha \log n + \beta) - 1)$ as correcting term. The estimate is now well centred but does not include the periodic terms, which are shown in Fig. 3.7. As we can see in Fig. 3.7 the fluctuation confirms the Conjecture 3.1 about the variance of Joint Complexity.

3.7   Numerical Experiments in Twitter

In this section we apply Joint Complexity in Twitter in order to perform topic detection and extract similarities between tweets and communities. We consider four sets of tweets from different sources—the New York Times, BBC Business,

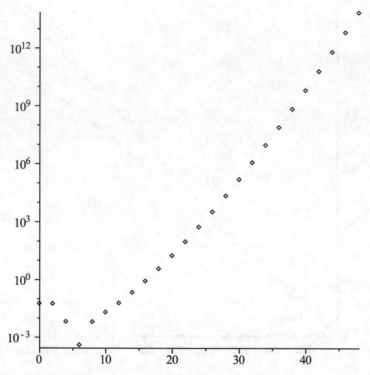

Fig. 3.5 Relative error in the saddle point expansion versus order for $x = 1/\beta$

CNN, and BBC Breaking. In Twitter the maximum length of the messages is 140 characters. We take the hypothesis that the sources are Markov sources of finite order. Individual tweets are of arbitrary length. The alphabets of the different languages of tweet sets are converted on ASCII.

We compute the JC value for pairs of tweet sets in Fig. 3.8. We used tweets from the 2012 Olympic Games and 2012 United States elections. We took two sets from each of these tweet sets to run our experiments, but first we removed the tags similar to the topic, such as #elections, #USelections, #USelections2012, #Olympics, #Olympics2012, #OlympicGames and so on. As we can see in Fig. 3.8, the JC is significantly high when we compare tweets in the same subjects, for both real and simulated tweets (simulated tweets are generated from a Markov source of order 3 trained on the real tweets). We observe the opposite when we compare different subjects. In the US elections topic, we can see that the JC increases significantly when the number of characters is between 1700 and 1900. This is because users begin to write about and discuss the same subject. We can observe the same in the Olympic Games topic between 6100 and 6300 and between 9500 and 9900. This shows the applicability of the method to distinguish information sources. In Fig. 3.9, we plot the JC between the simulated texts and compare with the theoretical average curves expected by the proposed methodology.

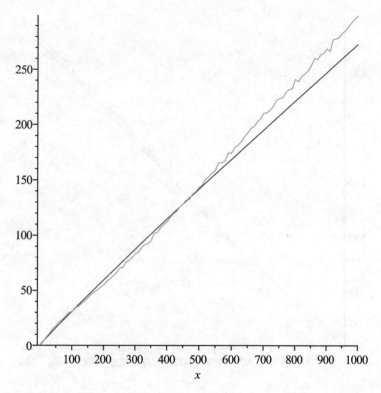

Fig. 3.6 Joint Complexity (*y* axis) of memoryless English text (*x* axis) versus monkey typing. The optimal fifth order theoretical average is shown in *red* (*cont. line*)

3.8   Suffix Trees

A Suffix Tree [93] is a compressed trie [77] containing all the suffixes of the given text as their keys and positions in the text as their values. We may refer as PAT tree or, in an earlier form, position tree. The suffix tree allows particularly fast implementations of many important string operations.

The construction of such a tree for a string S of length n takes on average $O(n \log n)$ time and space linear in the length of n. Once constructed, several operations can be performed quickly, for instance locating a substring in S, locating a substring if a certain number of mistakes are allowed, locating matches for a regular expression pattern and other useful operations.

Every node has outgoing edges labeled by symbols in the alphabet of S. Thus every node in the Suffix Tree can be identified via the word made of the sequence of labels from the root to the node. The Suffix Tree of S is the set of the nodes which are identified by any factor of S.

Suffix Tree Compression: A unitary sequence of nodes is a chain where nodes have all degree 1. If a unitary chain ends to a leaf then it corresponds to a factor

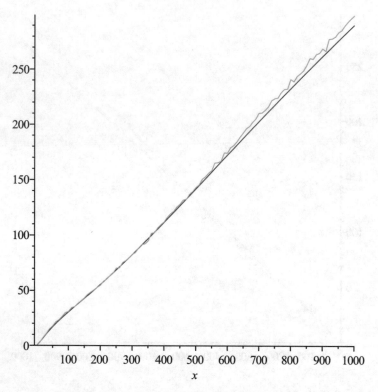

Fig. 3.7 Average Joint Complexity (y axis) of memoryless English text versus monkey typing. The optimal fifth order theoretical (in *red, cont. line*) plus periodic terms

which appears only once in S. The chain can be compressed into a single node. In this case the concatenation of all the labels of the chain correspond to a suffix of S and the compressed leaf will contain a pointer to this suffix. The other (internal) nodes of the Compressed Suffix Tree correspond to the factors which appear at least twice in S. This is the Compressed Suffix Tree version whose size is $O(n)$ in average, otherwise the uncompressed version is $O(n^2)$.

Similarly any other unitary chain which does not go to a leaf can also be compressed in a single node, the label of the edges to this node is the factor obtained by concatenating all the labels. This is called the Patricia compression and in general gives very small reduction in size.

The Suffix Tree implementation and the comparison process (ST superposition) between two Suffix Trees in order to extract the common factors of the text sequences can be found in the Appendix A.

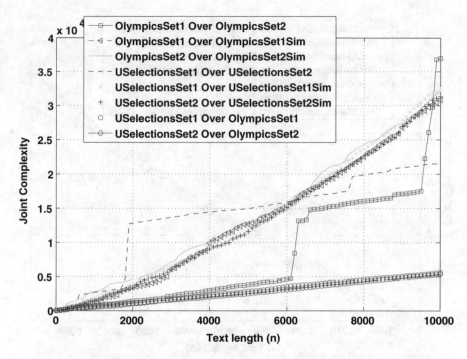

Fig. 3.8 Joint Complexity of four tweet sets from the 2012 United States elections and Olympic Games. The text is an incremental aggregation of tweets from these sets

3.8.1   Examples of Suffix Trees

The proposed tree structure in the Appendix A needs $O(n)$ time to be stored and sub linear time for the superposition (finding overlaps). Two main examples with graph tree representation follow in Figs. 3.10 and 3.11 for the sequence "apple" and "maple", respectively, which have nine common factors. Figures 3.12 and 3.13 show the Suffix Tree for the sequence "healthy" and "sealed", which have seven common factors. The construction of the Suffix Tree for the sequence "apple" and "maple", as well as the comparison between them (ST superposition) is shown in Fig. 3.14.

3.9   Snow Data Challenge

In February 2014 the Snow Data Challenge of the World Wide Web Conference (WWW'14) was announced. Every year the WWW community organizes a different challenge. In 2014 the challenge was about extracting topics in Twitter. The volume of information in Twitter is very high and it is often difficult to extract topics in real

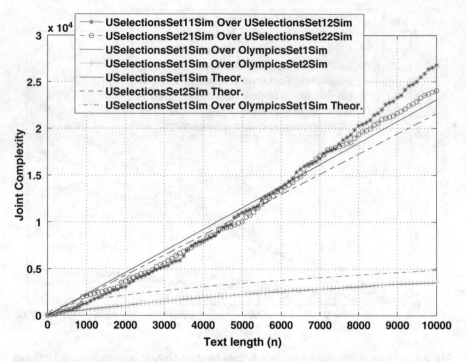

Fig. 3.9 Joint Complexity of tweet sets from the 2012 United States elections and Olympic games, in comparison with theoretic average curves. The text is an incremental aggregation of tweets from these sets

Fig. 3.10 Suffix Tree for the sequence "apple", where v is the empty string

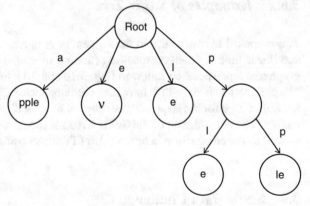

time. The task of this challenge was to automatically mine social streams to provide journalists a set of headlines and complementary information that summarize the most important topics for a number of timeslots (time intervals) of interest.

The Snow Challenge organization provided a common framework to mine the Twitter stream and asked to automatically extract topics corresponding to known

Fig. 3.11 Suffix Tree for the sequence "maple", where v is the empty string

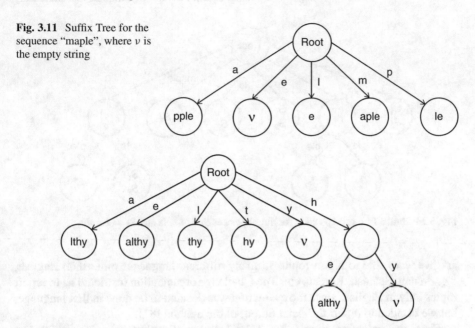

Fig. 3.12 Suffix Tree for the sequence "healthy", where v is the empty string

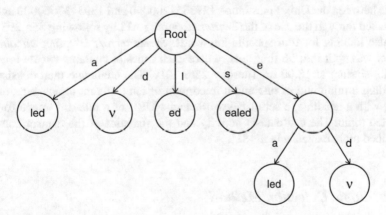

Fig. 3.13 Suffix Tree for the sequence "sealed", where v is the empty string

events (e.g. politics, sports, entertainment). The crawled data were divided into timeslots and we had to produce a fixed number of topics for selected timeslots. Each topic should be in the form of a short headline that summarizes a topic related to a piece of news occurring during that timeslot, accompanied by a set of tweets, URLs of pictures (extracted from the tweets), and a set of keywords. The expected output format was the following: *[headline, keywords, tweetIds, picture urls]*.

We got the third Prize in the Challenge, while our method was discussed to receive the first Prize. The main advantage of the method was its language agnostics

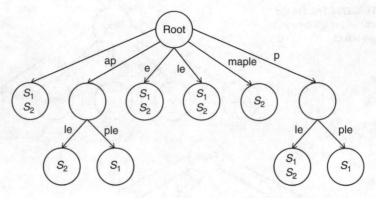

Fig. 3.14 Suffix Tree superposition for the sequences $S_1 = apple$ and $S_2 = maple$

and we were able to report topics in many different languages other than English, e.g. French, Spanish, Korean, etc. The Challenge organization restricted us to report topics only in English, since the evaluation was decided to be done in that language, but we decided to report the exact output of the method [81].

First, we collected tweets for 24 h; between Tuesday Feb. 25, 18:00 and Wednesday Feb. 26, 18:00 (GMT). The crawling collected more than 1,041,062 tweets between the Unix timestamps 1393351200000 and 1393437600000 and was conducted through the use of the *Twitter* streaming API by following 556,295 users and also looking for four specific keywords: *Syria; terror; Ukraine; bitcoin*. The dataset was split into 96 timeslots, where each timeslot contains tweets for every 15 min, starting at 18:00 on Tuesday 25th 2014. The challenge then consisted in providing a minimum of one and a maximum of ten different topics per timeslot, along with a headline, a set of keywords and a URL of a relevant image for each detected topic. The test dataset activity and the statistics of the dataset crawl are described more extensively in [81].

3.9.1   Topic Detection Method

Until the present, the main methods used for text classification are based on *keywords* detection and machine learning techniques as was extensively described in Chap. 2. Using keywords in tweets has several drawbacks because of wrong spelling or distorted usage of the words—it also requires lists of stop-words for every language to be built—or because of implicit references to previous texts or messages. The machine learning techniques are generally heavy and complex and therefore may not be good candidates for real-time text processing, especially in the case of Twitter where we have natural language and thousands of tweets per second to process. Furthermore, machine learning processes have to be manually initiated by tuning parameters, and it is one of the main drawbacks for the kind of application,

Fig. 3.15 Representation of
the first row of the n-th
timeslot via weighted graph

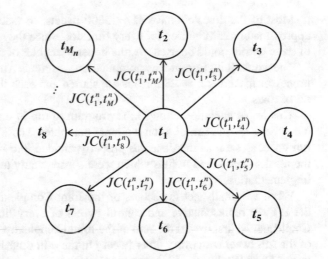

where we want minimum if any human intervention. Some other methods are using
information extracted by visiting the specific URLs on the text, which makes them
a heavy procedure, since one may have limited or no access to the information, e.g.
because of access rights, or data size and throughput.

In our method [9] we use the Joint Complexity (computed via Suffix Trees) as a
metric to quantify the similarity between the tweets. This is a significant achieve-
ment because we used a general method adapted to the Snow Data Challenge.

According to the dataset described in Sect. 3.9 and in [81] we have $N = 96$
timeslots with $n = 1 \ldots N$. For every tweet t_i^n, where $i = 1 \ldots M_n$, with M_n being
the total number of tweets, in the n-th timeslot, we build a Suffix Tree, $ST(t_i^n)$, as
described in Sect. 3.3. Building a Suffix Tree is an operation that costs linear time
and takes $O(m)$ space in memory, where m is the length of the tweet.

Then we compute the Joint Complexity metric as mentioned earlier, $JC(t_i^n, t_j^n)$ of
the tweet t_i^n with every other tweet t_j^n of the n-th timeslot, where $j = 1 \ldots M_n$, and
$j \neq i$ (by convention we choose $JC(t_i^n, t_i^n) = 0$). For the N timeslots we store the
results of the computation in the matrices $T_1, T_2, \ldots, T_N$ of $M_n \times M_n$ dimensions.

We represent each matrix T_n by fully connected weighted graphs. Each tweet is a
node in the graph and the two-dimensional array T_n holds the weight of each edge,
as shown in Fig. 3.15. Then, we calculate the score for each node in our graph by
summing all the edges which are connected to the node. The node that gives the
highest score is the most representative and central tweet of the timeslot.

$$
\mathbf{T}_n = \begin{pmatrix}
0 & JC(t_1^n, t_2^n) & JC(t_1^n, t_3^n) & \cdots & JC(t_1^n, t_M^n) \\
JC(t_2^n, t_1^n) & 0 & JC(t_2^n, t_3^n) & \cdots & JC(t_2^n, t_M^n) \\
JC(t_3^n, t_1^n) & JC(t_3^n, t_2^n) & 0 & \cdots & JC(t_3^n, t_M^n) \\
\vdots & \vdots & & \ddots & \vdots \\
JC(t_M^n, t_1^n) & JC(t_M^n, t_2^n) & JC(t_M^n, t_3^n) & \cdots & 0
\end{pmatrix}
$$

Most of the timeslots have $M = 5000$ tweets, so matrices $T_1, T_2, \ldots, T_N$ have approximately 25M entries for every timeslot. Since they are symmetric, only half of these entries could be used, i.e. the upper triangular of matrices $T_1, T_2, \ldots, T_N$.

The whole Joint Complexity computation was run in a multithreaded way on a 24 processor machine: $k = 23$ threads are started and each thread works on a disjoint set of rows.

This implementation allowed the program to run in on average 90 s in order to compute a 15-min timeslot, on a simple computer. These computations were only run once, as soon as the data was properly divided into the 15-min timeslots, and the results were saved in files which were subsequently used to perform the rest of implementation.

When we finally get the scores of the Joint Complexity metric, we try to find the R most representative and central tweets of every timeslot as required in the Challenge. At first we get the sum of the Joint Complexity, $S_i^n = \sum_{j=1\ldots M, j \neq i} JC_{t_i^n, t_j^n}$, of the i-th tweet with every other tweet j in the n-th timeslot, and finally we get the vector $\mathbf{S}^n = [S_1^n, S_2^n, \ldots, S_{M_n}^n]$ for every timeslot.

We sort the elements of each vector $\mathbf{S}^n$ in descending order and we get the R most representative and central tweets in the following way: The best-ranked tweet is chosen unconditionally, the second one is picked only if its JC score with the first one is below a chosen threshold Thr_{low}, otherwise it is added to the list of related tweets of the first tweet; similarly, the third one is picked only if its JC score with the first two is below Thr_{low}, etc. This ensures that the topics are dissimilar enough and it classifies best ranked tweets into topics at the same time.

3.9.2 Headlines

In order to produce *headlines* as requested from the Data Challenge, we removed punctuation, special characters, etc., of each selected tweet. We could use a better selection but due to the lack of time in preparation for the Challenge we used this simple selection method. We constructed the headlines of each topic and we run through the list of related tweets to keep only tweets that are different enough from the selected one (ensures no duplicates), we did so by keeping only the tweets whose JC score with the selected tweet and all previous related tweets was above a chosen threshold Thr_{max}. We first chose empirical values 400 and 600 for Thr_{low} and Thr_{max} respectively, but then we noticed that many topics had only one related tweet (all the others were retweets), so we decided to lower that threshold to $Thr_{low} = 240$. Due to the lack of time during the contest, we did not recompute the results on the whole dataset so only a handful of timeslots benefited from this better Thr_{low}. The final plan was to come up with a formula to have the system determine those thresholds automatically depending on the number of characters of each tweet.

While running through the list of related tweets we computed the *bag-of-words* used to construct the list of keywords and we also checked the original *.json* data to find a URL pointing to a valid image related to the topic.

We chose to print the first top eight topics for each timeslot, which are the heads of the first eight lists of related tweets.

3.9.3   Keywords Extraction

In order to produce a list of *keywords* per topic as requested from the Data Challenge, we first removed articles (stop-words), punctuation, special characters, etc., from the *bag-of-words* constructed from the list of related tweets of each topic. We got a list of words, and then we ordered them by decreasing frequency of occurrence. Finally, we reported the k most frequent words, in a list of keywords $\mathbf{K} = [K^1_{1\ldots k}, K^2_{1\ldots k}, \ldots, K^N_{1\ldots k}]$, for the N total number of timeslots.

3.9.4   Media URLs

As requested from the Data Challenge, we provided a representative Media URL per topic. The body of a tweet (in the *.json* file format) contains a URL information for links to media files such as pictures or videos, when available this information is stored in the following subsection of the *json* object: *entities* $\rightarrow$ *media* $\rightarrow$ *media_url*. While reporting the most representative and central tweets, we scan the original json format in order to retrieve such a URL, from the most representative tweet or any of its related tweets, pointing to valid photos or pictures in a *.jpg*, *.png* or *.gif* format. Then, we report these pictures along with the headlines and the set of keywords, as shown in Algorithm 1.

Almost half of the headlines (47%) produced by our method had a picture retrieved in the *.json* file. When no such URL is provided within the collected *json* objects we planned to visit the URLs of websites and retrieve images, which were online, in a way that were enforced to be relevant for the topic, but we did not follow that strategy because of the study of the Google API and due to lack of time. It is important to point out that the goal of the challenge was to detect newsworthy items before they hit mainstream news websites, so it was decided that parsing images from such websites was not interesting in that context.

Algorithm 1 Topic detection based on Joint Complexity

// N = # timeslots, M = # tweets in the n-th timeslot
for $n = 1$ *to* N **do**
    for $t = 1$ *to* M **do**
        $t \leftarrow t_{json}.getText()$;
        $t_{ST} \leftarrow suffixTreeConstruction(t)$;
        $JCScores \leftarrow JCMetric()$;
    end for
    // Find the most representative & central tweets
    $S^n \leftarrow sum(JCScores)$;
    // Get headlines for the central tweets
    $R^n \leftarrow descendingOrder(S^n)$;
    // Get set of keywords
    $K^n \leftarrow keywords(R^n)$;
    // Get URLs of pictures from the .json file
    $P^n \leftarrow mediaURL(R^n)$;
    // Print the results in appropriate format
    $Print(R^n)$;
    $Print(K^n)$;
    $Print(P^n)$;
end for

3.9.5 Evaluation of Topic Detection

Apart from the specific implementation for the Snow Data Challenge, the main benefits of our method are that we can both classify the messages and identify the growing trends in real time, without having to manually set up lists of keywords for every language. We can track the information and timelines within a social network and find groups of users which agree on the same topics.

The official evaluation results of our method in the Snow Data Challenge are included in [81]. Although the dataset that was used for this challenge did not allow to show this properly, one key advantage of using Joint Complexity is that it can deal with languages other than English [46, 66] without requiring any additional feature.

3.10 Tweet Classification

3.10.1 Tweet Augmentation

Joint Complexity was also used for classification in Twitter [65]. The innovation brought by the method is in the use of the information contained in the redirected URLs of tweets. We use this information to augment the similarity measure of

JC, which we call *tweet augmentation*. It must be noted that this method does not have access to the redirected URLs as described above about the prior art existing solution.

The method proceeds in two phases: (3.10.2) Training phase, and (3.10.3) Run phase.

3.10.2 Training Phase

During the *Training phase* we construct the training databases (DBs) by using Twitter's streaming API with filters for specific keywords. For example, if we want to build a class about politics, then we ask the Twitter API for tweets that contain the word "politics". Using these requests we build M classes on different topics. Assume that each class contains N tweets (e.g. $M = 5$ Classes: politics, economics, sports, technology, lifestyle of $N = 5000$ tweets). To each class we allocate K keywords (e.g. the keywords used to populate the class; their set is smaller than the *bag-of-words*). The tweets come in the *.json* format which is the basic format delivered by the Twitter API.

Then we proceed to the URL extraction and tweet augmentation. The body of a tweet (in the *.json* file format) contains a URL information if the original author of the tweet has inserted one. In general Twitter applies a hashing code in order to reduce the link size in the tweets delivered to users (this is called *URL shortening*). However the original URL comes in clear in the *.json* format provided by the Twitter API. While extracting the tweet itself, we get both the hashed URL and the original URL posted by the user. Then, we replace the short URL in the tweet's text by the original URL and we get the augmented tweet.

In the next step, we proceed with the Suffix Tree construction of the augmented tweet. Building a suffix tree is an operation that costs $O(n \log n)$ operations and takes $O(n)$ space in memory, where n is the length of the augmented tweet. The tweet itself does not exceed 140 characters, so the total length of the augmented tweet is typically smaller than 200 characters.

3.10.3 Run Phase

During the *Run phase* (shown in Algorithm 2), we get tweets from the Twitter Streaming Sample API. For every incoming tweet we proceed to its classification by the following operations: At first, we augment the tweet as described in Sect. 3.10.2 Training phase. Then, we compute the matching metric of the augmented incoming tweet with each class. The score metric is of the form:

$$MJC * \alpha + PM * \beta \tag{3.19}$$

Algorithm 2 Tweet classification based on Joint Complexity and pattern matching

---

Training phase:

> $constructClasses(M, N)$;
>
> **for** $i = 1$ *to* M **do**
> > **for** $i = 1$ *to* N **do**
> > > $t_{i,j}^{URL} \leftarrow extractURL(t_{ij})$;
> > >
> > > $t_{i,j}^{aug} \leftarrow tweetAugmentation(t_{ij}, t_{i,j}^{URL})$;
> > >
> > > $t_{i,j}^{ST} \leftarrow suffixTreeConstruction(t_{ij}^{aug})$;
> >
> > **end for**
> **end for**

Run phase:

> **while** $(t_{json} \leftarrow TwitterAPI.getSample() \mathrel{!}= null)$ **do**
> > $t \leftarrow t_{json}.getText()$;
> >
> > $t_{URL} \leftarrow extractURL(t_{json})$;
> >
> > $t_{aug} \leftarrow tweetAugmentation(t, t_{URL})$;
> >
> > $t_{ST} \leftarrow suffixTreeConstruction(t_{aug})$;
> >
> > **for** $i = 1$ *to* M **do**
> > > $PM_i(t) \leftarrow patternMatching(t_{URL})$;
> > >
> > > $JC_i^{avg} \leftarrow averageJC(t_{aug})$;
> > >
> > > $JC_i^{max} \leftarrow maximum(JC(t_{aug}))$;
> > >
> > > $\beta \leftarrow \dfrac{JC_i^{max} - JC_i^{avg}}{JC_i^{max}}$
> > >
> > > $\alpha \leftarrow 1 - \beta$
> >
> > **end for**
> >
> > $D(t) \leftarrow arg\max_{i=1}^{M}\{\max_{j=1}^{N}(JC(t_{i,j}, t) * \alpha) + PM_i * \beta\}$
> >
> > $classifyTweet(t, D(t))$;
>
> **end while**
>
> $UpdateByDeletingOldestTweets()$;

---

where *MJC* is the max of Joint Complexity (JC) of the augmented incoming tweet over the tweets already present in the class, and *PM* is the pattern matching score of the incoming tweet over the class keywords. Quantities α and β are weight parameters, which depend on the average Joint Complexity, JC_i^{avg}, of the i-th class, and the maximum JC (best fitted), JC_i^{max}. We construct those as follows:

$$\beta = \frac{JC_i^{max} - JC_i^{avg}}{JC_i^{max}}$$

$$\alpha = 1 - \beta$$

When the average Joint Complexity, $JC_i^{avg} = \frac{JC_i^{max}}{2}$ the weight $\alpha = \beta = 0.5$, and if the pattern matching on the URL returns zero, then $\beta = 0$ and $\alpha = 1$.

The Joint Complexity between two tweets is the number of the common factors defined in language theory and can be computed efficiently in $O(n)$ operations (sublinear on average) by Suffix Tree superposition. We also compute the Pattern Matching score with the keywords of the i-th class, i.e. as the number of keywords actually present in the augmented tweet URL. The metric is a combination of the *MJC* and *PM*.

We then assign an incoming tweet to the class that maximizes the matching metric defined at (3.19) and we also link it to the best fitted tweet in this class, i.e. the tweet that maximizes the Joint Complexity inside this class.

In the case described above where newly classified tweets are added to the reference class (which is useful for *trend sensing*), then in order to limit the size of each reference class we delete the oldest tweets or the least significant ones (e.g. the ones which got the lowest JC score). This ensures the low cost and efficiency of our method.

The main benefits of our method are that we can both classify the messages and identify the growing trends in real time, without having to manually identify lists of keywords for every language. We can track the information and timeline within a social network and find groups of users that agree or have the same interests, i.e, perform trend sensing.

3.10.4 *Experimental Results on Tweet Classification*

The efficiency of the proposed classification method is evaluated on sets of tweets acquired from the Twitter API. The classification accuracy of five tested methods was measured with the standard *Precision*, *Recall* and *F-score* metrics (detailed in the section below), using a Ground Truth (GT) shown in Fig. 3.16. The experiments were run on more than 1M tweets [81].

Instead of retrieving live tweets from the Twitter Streaming API as described in Algorithm 2, we stored a list of random tweet IDs in a file so that each algorithm would work on the same set of tweets.

We selected the Document-Pivot (DP) method to compare with our new method, since it outperformed most of the other state-of-the-art techniques in a Twitter context as shown in [2].

In order to run the experiment, we modified an existing implementation of Document-Pivot which was developed for topic detection in [2]. The modification consisted in setting up almost the same Training phase as for the Joint Complexity implementation (i.e. use the same reference tweets), with a notable exception for the *tweet augmentation*. For the latter implementation, instead of placing the original URL, we first decompose the URL into a list of pseudo-words by replacing the "/" character by a space. This method, named DPurl, will prove useful for the classification as quite often URLs contain specific keywords such as *sport*, *politics*, etc.

Fig. 3.16 Ground Truth
distribution of the tweets into
the different categories

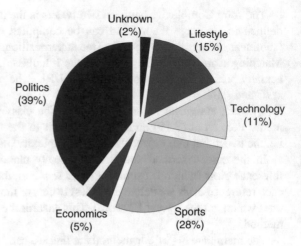

The other and more important difference with DP (without changing the output of
the method) is that instead of building Suffix Trees, this time the method constructs
a *tf-idf* bag of words, and then classifies each tweet of the Run phase by selecting
the category containing the *closest* tweet to our test tweet. The notion of *closest*
is because we used Locality Sensitive Hashing based on the Cosine Similarity in
a vector space where each possible word is a dimension and its *tf-idf* score is the
coordinate in that dimension. In such a space when the cosine between the two
vectors is close to 1, it means that the vectors are pointing in the roughly same
direction, in other words the two tweets represented by the vectors should share a
lot of words and thus should probably speak about or refer to the same subject.

3.10.4.1   Classification Performance Based on Ground Truth

The classification performance is compared for five methods, which are:

1. Document Pivot (DP), without tweet augmentation,
2. Joint Complexity (JC), without tweet augmentation,
3. Document Pivot with URL (DPurl) described above,
4. Joint Complexity with URL (JCurl) described above, without the Pattern Matching,
5. Joint Complexity with URL and Pattern Matching (JCurlPM) described in
   Algorithm 2 in Sect. 3.10.

The standard *Precision*, *Recall* and *F-score* metrics were used to evaluate the
different classification output, they are described below:

$$Precision = \frac{true\ positives}{true\ positives + false\ positives}$$

Table 3.2 Average of
precision, recall and *F-score*
for the used classification
methods for all classes

Metric	DP	JC	DPurl	JCurl	JCurlPM
Precision	0.47	**0.60**	0.68	**0.71**	**0.86**
Recall	0.38	**0.48**	0.57	**0.63**	**0.86**
F-score	0.42	**0.53**	0.62	**0.67**	**0.86**

The methods DP, JC, Purl, JCurl and JCurlPM are used

$$Recall = \frac{true\ positives}{true\ positives + false\ negatives}$$

where, for a class $\mathscr{C}$, *true positives* are tweets that were classified in $\mathscr{C}$ by both the algorithm and the Ground Truth, *false positives* are tweets that were classified in $\mathscr{C}$ by the algorithm but in some other class by the Ground Truth and *false negatives* are tweets that were classified in $\mathscr{C}$ by the Ground Truth but in some other class by the algorithm.

We also computed the *F-score* in order to combine into a single metric both precision and recall (for faster comparison at a glance):

$$F\text{-}score = 2 * \frac{Recall * Precision}{Recall + Precision}$$

A global overview of the results is presented in Table 3.2 where we can see that, on average, JC outperforms DP, JCurl outperforms DPurl and JCurlPM clearly outperforms them all.

Looking in more details for each category, the global tendency is confirmed except for a couple of categories like *Technology* where DP has a slightly better Precision but a worse Recall. In the *Sports* category, on the other hand, the situation is reversed as DP seems to provide a slightly better Recall. In both cases the differences are too small to be really significant and what can be noted is that JCurlPM always outperforms all other methods (Fig. 3.17). The mediocre precision obtained by DP and JC in Fig. 3.18 can be explained by the fact that the *Economics* category was under-represented in the Ground Truth dataset and given the fact that *Politics* and *Economics* are often very close subjects, both methods classified a few *Politics* tweets into the *Economics* category thus lowering the Precision. It can be noted that the Recall, on the other hand, is quite good for both methods (Figs. 3.19, 3.20 and 3.21).

3.11   Chapter Summary

In this chapter we studied the Joint Sequence Complexity and its applications, which range from finding similarities between sequences to source discrimination. Markov models well described the generation of natural text, and we exploited datasets from different natural languages using both short and long sequences. We provided

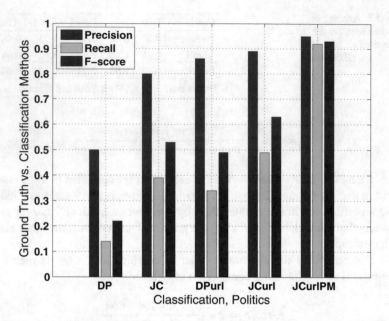

Fig. 3.17 Precision (*left*), recall (*middle*) and F-score (*right*) for the classified tweets in the class Politics

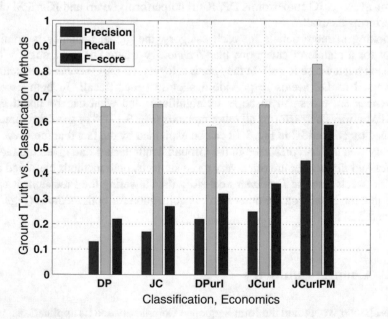

Fig. 3.18 Precision (*left*), recall (*middle*) and F-score (*right*) for the classified tweets in the class Economics

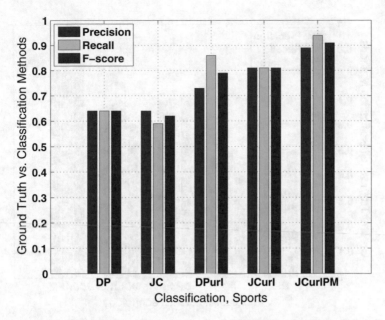

Fig. 3.19 Precision (*left*), recall (*middle*) and F-score (*right*) for the classified tweets in the class Sports

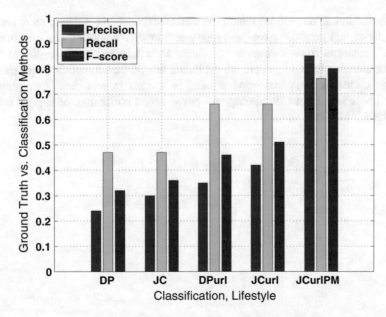

Fig. 3.20 Precision (*left*), recall (*middle*) and F-score (*right*) for the classified tweets in the class Lifestyle

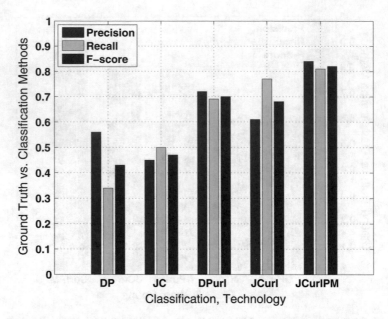

Fig. 3.21 Precision (*left*), recall (*middle*) and F-score (*right*) for the classified tweets in the class Technology

models and notations, and presented the theoretical analysis. A study on expending asymptotics and periodic terms was also mentioned. We applied our methodology to real messages from Twitter in the Snow Data Challenge of the World Wide Web Conference in 2014, where we evaluated our proposed methodology on topic detection, classification and trend sensing in Twitter in real time. Our proposed method based on Joint Complexity was praised by a committee of experts and we won the third Prize.

Chapter 4
Text Classification via Compressive Sensing

Abstract In this chapter we apply the theory of Compressive Sensing (CS) to achieve low dimensional classification. According to Compressive Sensing theory, signals that are sparse or compressible in a suitable transform basis can be recovered from a highly reduced number of incoherent linear random projections, which overcomes the traditional signal processing methods. Traditional methods are dominated by the well-established Nyquist–Shannon sampling theorem, which requires the sampling rate to be at least twice the maximum bandwidth. We introduce a hybrid classification and tracking method, which extends our recently introduced Joint Complexity method, which was tailored to the topic detection and trend sensing of user's tweets. First we employ the Joint Complexity, already described in detail in the previous chapter to perform topic detection, and then, based on the nature of the data, we apply the methodology of Compressive Sensing to perform topic classification by recovering an indicator vector. Finally, we combine the Kalman filter, as a refinement step for the update of the tracking process.

4.1 Introduction

In this chapter we apply the theory of Compressive Sensing (CS) [18] to achieve low dimensional classification. According to Compressive Sensing theory, signals that are sparse or compressible in a suitable transform basis can be recovered from a highly reduced number of incoherent linear random projections, which overcomes the traditional signal processing methods. Traditional methods are dominated by the well-established Nyquist–Shannon sampling theorem, which requires the sampling rate to be at least twice the maximum bandwidth.

We introduce a hybrid classification and tracking method, which extends our recently introduced Joint Complexity method [46, 66], which was tailored to the topic detection and trend sensing of user's tweets. More specifically, we propose a two-step detection, classification and tracking method:

First we employ the Joint Complexity, already described in detail in the previous chapter, as the cardinality of a set of all distinct factors of a given string represented by suffix trees, to perform topic detection. Second, based on the nature of the data, we apply the methodology of Compressive Sensing to perform topic classification

© Springer International Publishing AG 2018

D. Milioris, *Topic Detection and Classification in Social Networks*,

DOI 10.1007/978-3-319-66414-9_4

by recovering an indicator vector. Finally, we combine the Kalman filter, as a refinement step for the update of the tracking process.

4.2 Compressive Sensing Theory

Let us first describe the main theoretical concepts of CS [4, 12, 18] and how it is applied on the problem classification [61, 68]. Consider a discrete-time signal x in $\mathbb{R}^N$. Such signal can be represented as a linear combination of a set of basis $\{\psi_i\}_{i=1}^N$. Constructing an $N \times N$ basis matrix $\boldsymbol{\Psi} = [\psi_1, \psi_2, \ldots, \psi_N]$, the signal x can be expressed as

$$x = \sum_{i=1}^N s_i \psi_i = \boldsymbol{\Psi} s \qquad (4.1)$$

where $s = (s_1, s_2, \ldots s_N) \in \mathbb{R}^N$ and is an equivalent representation of x in a basis $\boldsymbol{\Psi}$.
   In fact the signal is represented as

$$x = \boldsymbol{\Psi} s + \theta \qquad (4.2)$$

with $\theta \in \mathbb{R}^N$ being the noise, where $\mathbb{E}(\lessgtr) = 0$ and $var(\theta) = O(|\boldsymbol{\Psi} s|)$. The efficiency of a CS method for signal approximation or reconstruction depends highly on the sparsity structure of the signal in a suitable transform domain associated with an appropriate sparsifying basis $\boldsymbol{\Psi} \in \mathbb{R}^{N \times N}$. It has been demonstrated [12, 18] that if $\mathbf{x}$ is K-sparse in $\boldsymbol{\Psi}$ (meaning that the signal is exactly or approximately represented by K elements of this basis), it can be reconstructed from $M = rK \ll N$ non-adaptive linear projections onto a second measurement basis, which is incoherent with the sparsity basis, and where r is a small overmeasuring factor ($r > 1$).
   The measurement model in the original space-domain is expressed as $\mathbf{g} = \boldsymbol{\Phi} \mathbf{x}$, where $\mathbf{g} \in \mathbb{R}^M$ is the measurement vector and $\boldsymbol{\Phi} \in \mathbb{R}^{M \times N}$ denotes the measurement matrix. By noting that $\mathbf{x}$ can be expressed in terms of the basis $\boldsymbol{\Psi}$ as in (4.2) the measurement model has the following equivalent transform-domain representation

$$\mathbf{g} = \boldsymbol{\Phi} \boldsymbol{\Psi} \mathbf{s} + \boldsymbol{\Phi} \theta \ . \qquad (4.3)$$

In fact when the length of the sequence (i.e. tweet) $n \to \infty$ and $N \to \infty$, $\mathbb{E}(\boldsymbol{\Psi} s) = O(nN)$, with $var(\theta) = O(nN)$, $std(\theta) = O\left(\sqrt{|\boldsymbol{\Phi}|n}\right)$ and $\mathbb{E}(\boldsymbol{\Phi} \theta) = 0$. The second part of (4.3), $\boldsymbol{\Phi} \theta$ is of relative order $O\left(\frac{1}{\sqrt{nN}}\right)$, and is negligible compared to $\boldsymbol{\Phi} \boldsymbol{\Psi} \mathbf{s}$ due to the law of large numbers. Examples of measurement matrices $\boldsymbol{\Phi}$, which are incoherent with any fixed transform basis $\boldsymbol{\Psi}$ with high probability (universality property [18]), are random matrices with independent and identically distributed (i.i.d.) Gaussian or Bernoulli entries. Two matrices $\boldsymbol{\Psi}$, $\boldsymbol{\Phi}$ are incoherent if the elements of the first are not represented sparsely by the elements of the second, and vice versa. Since the original vectors of signals, $\mathbf{x}$, are not sparse in general, in the following study we focus on the more general case of reconstructing their equivalent sparse representations, $\mathbf{s}$, given a low-dimensional set of measurements $\mathbf{g}$ and the measurement matrix $\boldsymbol{\Phi}$.

By employing the M compressive measurements and given the K-sparsity property in basis $\boldsymbol{\Psi}$, the sparse vector $\mathbf{s}$, and consequently the original signal $\mathbf{x}$, can be recovered perfectly with high probability by taking a number of different approaches. In the case of noiseless CS measurements the sparse vector $\mathbf{s}$ is estimated by solving a constrained ℓ_0-norm optimization problem of the form,

$$\hat{\mathbf{s}} = \arg\min_{\mathbf{s}} \|\mathbf{s}\|_0, \quad \text{s.t.} \quad \mathbf{g} = \boldsymbol{\Phi}\boldsymbol{\Psi}\mathbf{s}, \tag{4.4}$$

where $\|\mathbf{s}\|_0$ denotes the ℓ_0 norm of the vector $\mathbf{s}$, which is defined as the number of its non-zero components. However, it has been proven that this is an NP-complete problem, and the optimization problem can be solved in practice by means of a relaxation process that replaces the ℓ_0 with the ℓ_1 norm,

$$\hat{\mathbf{s}} = \arg\min_{\mathbf{s}} \|\mathbf{s}\|_1, \quad \text{s.t.} \quad \mathbf{g} = \boldsymbol{\Phi}\boldsymbol{\Psi}\mathbf{s}. \tag{4.5}$$

which will give s with a relative error of $O\left(\frac{1}{\sqrt{nN}}\right)$. In [12, 18] it was shown that these two problems are equivalent when certain conditions are satisfied by the two matrices $\boldsymbol{\Phi}, \boldsymbol{\Psi}$ (restricted isometry property (RIP)).

The objective function and the constraint in (4.5) can be combined into a single objective function, and several of the most commonly used CS reconstruction methods solve the following problem,

$$\hat{\mathbf{s}} = \arg\min_{\mathbf{s}} \left(\|\mathbf{s}\|_1 + \tau\|\mathbf{g} - (\boldsymbol{\Phi}\boldsymbol{\Psi}\mathbf{s})\|_2 \right), \tag{4.6}$$

where τ is a regularization factor that controls the trade-off between the achieved sparsity (first term in (4.6)) and the reconstruction error (second term). Commonly used algorithms are based on linear programming [14], convex relaxation [12, 96] and greedy strategies (e.g. Orthogonal Matching Pursuit (OMP) [19, 98]).

4.3 Compressive Sensing Classification

4.3.1 Training Phase

During the training phase, we built our classes as described in Sect. 3.3 and for each class we extract the most central/representative tweet(s) (CTs) based on the Joint Complexity method. The vector $\boldsymbol{\Psi}_T^i$ consists of the highest JC scores of

Fig. 4.1 Flowchart of the
preprocessing phase of Joint
Complexity

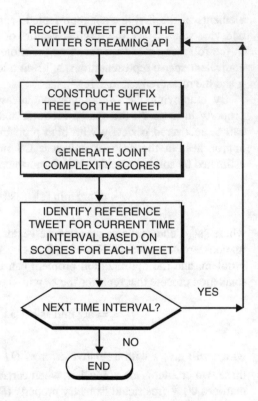

the i-th CT. The matrix $\boldsymbol{\Psi}_T$ is used as the appropriate sparsifying dictionary for
the training phase. Moreover, a measurement matrix $\boldsymbol{\Phi}_T^i$ is associated with each
transform matrix $\boldsymbol{\Psi}_T^i$. In the proposed algorithm, a standard Gaussian measurement
matrix is employed, with its columns being normalized to unit ℓ_2 norm (Fig. 4.1).

Figure 4.2 shows a flowchart of the preprocessing phase classification based on
Compressive Sensing in conjunction with the part of Joint Complexity shown in the
flowchart of Fig. 4.1.

4.3.2   Run Phase

A similar process is followed during the runtime phase. More specifically, we denote
$\mathbf{x}_{c,R}$ as the Joint Complexity score of the incoming tweet with the CT_i classified at the
current class c, where R denotes the runtime phase. The runtime CS measurement
model is written as

$$\mathbf{g}_c = \boldsymbol{\Phi}_R \mathbf{x}_{c,R} , \qquad (4.7)$$

where $\boldsymbol{\Phi}_R$ denotes the corresponding measurement matrix during the runtime phase.

Fig. 4.2 Flowchart of the
preprocessing phase of
Compressive Sensing

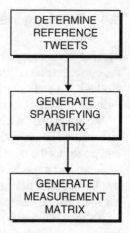

Fig. 4.3 Flowchart of the
runtime phase of the
Compressive Sensing based
classification

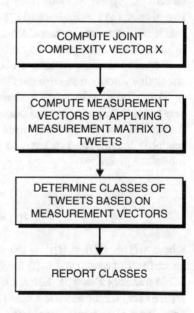

The measurement vector $\mathbf{g}_c$ is formed for each CT_i according to (4.7) and the reconstruction takes place via the solution of (4.6), with the training matrix $\boldsymbol{\Psi}_T$ being used as the appropriate sparsifying dictionary.

Figure 4.3 shows the flowchart for the runtime phase of the classification based on Compressive Sensing.

In this work, we are based on the assumption that the CS-based classification method involves the mobile device that collects the tweets from the Twitter API and performs the core CS algorithm. The performance analysis, described in Sect. 4.5, reveals an increased accuracy of the proposed CS-based classification algorithm when compared with other methods described in Sects. 2.4 and 4.5.

4.4 Tracking via Kalman Filter

Most of the tracking methods use past state estimates and motion dynamics to refine the current state estimate determined by the above topic detection and classification methods. In addition, the dynamic motion model can also be used in conjunction with the current state estimate to predict the future possible states.

We are based on the assumption that the Compressive Sensing based classification method involves the mobile device that collects the tweets from the Twitter API and performs the core Joint Complexity and Compressive Sensing algorithm.

If we had a model of Joint Complexity to detect the change of topics, we could use the Kalman filter to track a user according to his/her tweets. In this work, we assume that the change of topics is uniform.

Kalman filtering is a well-established method for estimating and tracking mobile targets. A typical Kalman filter [29] is applied recursively on a given dataset in two phases: (1) *Prediction* and (2) *Update*. The main advantage of this algorithm is that it can be executed in real time, since it is only based on the currently available information and the previously estimated state.

Focusing on the problem of classification, the user tweets periodically, and we check that information with the CTs at a specific time interval Δt.

Then, the classification system estimates the user's class at time t, which is denoted by $p^*(t) = [x^*(t)]^T$. Following a Kalman filtering approach, we assume that the process and observation noises are Gaussian, and also that the motion dynamics model is linear. The process and observation equations of a Kalman filter-based model are given by

$$\mathbf{x}(t) = \mathbf{F}\mathbf{x}(t-1) + \boldsymbol{\theta}(t) \tag{4.8}$$

$$\mathbf{z}(t) = \mathbf{H}\mathbf{x}(t) + \mathbf{v}(t) \tag{4.9}$$

where $\mathbf{x}(t) = [x(t), v_x(t)]^T$ is the state vector, with x being the correct class in the space (user's tweets) and $v_x(t)$ the tweeting frequency, $\mathbf{z}(t)$ is the observation vector, while matrices $\mathbf{F}$ and $\mathbf{H}$ define the linear motion model. The process noise $\boldsymbol{\theta}(t) \sim N(\mathbf{0}, \mathbf{S})$ and the observation noise $\mathbf{v}(t) \sim N(\mathbf{0}, \mathbf{U})$ are assumed to be independent zero-mean Gaussian vectors with covariance matrices $\mathbf{S}$ and $\mathbf{U}$, respectively. The current class of the user is assumed to be the previous one plus the information provided by the JC metric, which is computed as the time interval Δt multiplied by the current tweeting speed/frequency.

The steps to update the current estimate of the state vector $\mathbf{x}^*(t)$, as well as its error covariance $\mathbf{P}(t)$, during the prediction and update phase are given by the following equations

$$\mathbf{x}^{*-}(t) = \mathbf{F}\mathbf{x}^*(t-1) \tag{4.10}$$

$$\mathbf{P}^-(t) = \mathbf{F}\mathbf{P}(t-1)\mathbf{F}^\mathbf{T} + \mathbf{S} \tag{4.11}$$

$$\mathbf{K}(t) = \mathbf{P}^-(t)\mathbf{H}^T(\mathbf{H}\mathbf{P}^-(t)\mathbf{H}^T + \mathbf{U})^{-1} \tag{4.12}$$

$$\mathbf{x}^*(t) = \mathbf{x}^{*-}(t) + \mathbf{K}(t)(\mathbf{z}(t) - \mathbf{H}\mathbf{x}^{*-}(t)) \tag{4.13}$$

$$\mathbf{P}(t) = (\mathbf{I} - \mathbf{K}(t)\mathbf{H})\mathbf{P}^-(t) \tag{4.14}$$

where the superscript "$-$" denotes the prediction at time t, and $\mathbf{K}(t)$ is the optimal Kalman gain at time t.

The proposed Kalman system exploits not only the highly reduced set of compressed measurements, but also the previous user's class to restrict the classification set. The Kalman filter is applied on the CS-based classification [62], described briefly in Sect. 4.3, to improve the estimation accuracy of the mobile user's path. More specifically, let $\mathbf{s}^*$ be the reconstructed position-indicator vector. Of course in practice $\mathbf{s}^*$ will not be truly sparse, thus the current estimated position $[x_{CS}]$, or equivalently, cell c_{CS}, corresponds to the highest-amplitude index of $\mathbf{s}^*$. Then, this estimate is given as an input to the Kalman filter by assuming that it corresponds to the previous time $t - 1$, that is, $\mathbf{x}^*(t - 1) = [x_{CS}, v_x(t - 1)]^T$, and the current position is updated using (4.10). At this point, we would like to emphasize the computational efficiency of the proposed approach, since it is solely based on the use of the very low-dimensional set of compressed measurements given by (4.3), which are obtained via a simple matrix-vector multiplication with the original high-dimensional vector. Given the limited memory and bandwidth capabilities of a small mobile device, the proposed approach can be an effective candidate to achieve accurate information propagation, while increasing the device's lifetime. Since $M \ll N$ we have a great complexity improvement given by Compressive Sensing, which reduces the overall complexity of the Kalman filter. Algorithm 3 shows the combination of JC and CS method in conjunction with the Kalman filter, and summarizes the proposed information propagation system. Finally, Fig. 4.4 presents the flowchart for generating a tracking model and predicting classes of tweets.

Algorithm 3 Tweet classification based on JC, CS and Kalman

Training phase:

  // Build Classes according to JC
  constructClasses(M, N);

Run phase:

  while ($t_{json} \leftarrow$ *TwitterAPI.getSample*() ! $=$ *null*) **do**
    // Classify t by running the CS classification algorithm
    Give the estimated class as input to the Kalman filter
  end while
  Update by deleting oldest tweets

Fig. 4.4 Flowchart of the
classification and tracking
model based on Kalman filter
and Compressive Sensing

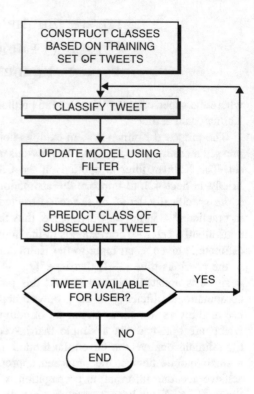

4.5 Experimental Results

The efficiency of the proposed classification method is evaluated on sets of tweets
acquired from the Twitter API. The classification accuracy of the tested methods
was measured with the standard *Precision*, *Recall* and *F-score* metrics as already
defined in Chap. 3, using a Ground Truth (GT) on more than 1M tweets [81].

The GT was computed by averaging the values returned by users and kept track
of which tweet was classified in which class in order to compare this with four
classification methods along with many different optimization techniques for the
signal reconstruction mentioned in Sect. 4.3.

We selected the Document-Pivot (DP) method to compare with our new method,
since it outperformed most of the other state-of-the-art techniques in a Twitter
context as shown in [2]. The most important difference of DP method is that instead
of building suffix trees, this time the method constructs a *tf-idf* bag of words), and
then classifies each tweet of the Run phase by selecting the category containing the
tweet *closest* to our test tweet. The notion of *closest* is because we used Locality
Sensitive Hashing based on the Cosine Similarity in a vector space where each
possible word is a dimension and its *tf-idf* score is the coordinate in that dimension.
In such a space when the cosine between the two vectors is close to 1, it means that
the vectors are pointing in the roughly same direction, in other words the two tweets
represented by the vectors should share a lot of words and thus should probably
speak about or refer to the same subject.

Table 4.1 Precision, recall and F-score for the used classification methods for all classes

Metric	DP	JC+CS	DPurl	JCurl+CS
Precision	0.47	**0.78**	0.68	**0.89**
Recall	0.38	**0.56**	0.57	**0.81**
F-score	0.42	**0.65**	0.62	**0.85**

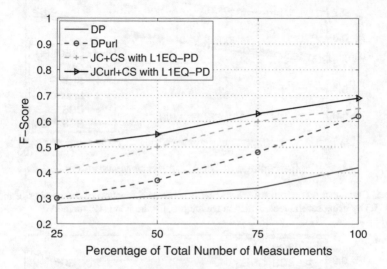

Fig. 4.5 Classification accuracy measured by F-Score for the DP, DPurl and JC+CS, JCurl+CS method as a function of the number of measurements (%) by using the ℓ_1-norm minimization

4.5.1 Classification Performance Based on Ground Truth

The classification performance is compared for: (a) Document Pivot (DP), (b) Joint Complexity with Compressive Sensing (JC+CS), (c) Document Pivot with URL (DPurl), (d) Joint Complexity and Compressive Sensing with URL (JCurl+CS), where (c) and (d) include the information of the compressed URL of a tweet concatenated with the original tweet's text and extracted from the *.json* file.

An overview of the results is presented in Table 4.1 where we can see that, on average, JC with CS outperforms DP, and JCurl with CS outperforms DPurl.

Figure 4.5 compares the classification accuracy of the DP, DPurl and JC+CS, JCurl+CS method as a function of the number of measurements by using the ℓ_1-norm minimization. Figure 4.6 compares the reconstruction performance between several widely used norm based techniques and Bayesian CS algorithms. More specifically, the following methods are employed[1]: (1) ℓ_1-norm minimization using the primal-dual interior point method (L1EQ-PD), (2) Orthogonal Matching Pursuit

---

[1]For the implementation of methods (1)–(5) the MATLAB codes can be found in: http://sparselab. stanford.edu/, http://www.acm.caltech.edu/l1magic, http://people.ee.duke.edu/~lcarin/BCS.html

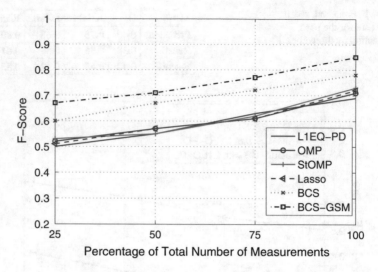

Fig. 4.6 Classification accuracy measured by F-Score as a function of the number of measurements (%) by using several reconstruction techniques, for the JCurl+CS method

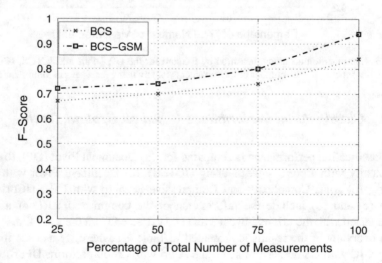

Fig. 4.7 Classification accuracy measured by F-Score as a function of the number of measurements (%) by using Kalman, for the JCurl+CS method, with the BCS and BCS-GSM reconstruction techniques

(OMP), (3) Stagewise Orthogonal Matching Pursuit (StOMP), (4) LASSO, (5) BCS and (6) BCS-GSM [100]. Figure 4.6 shows that BCS and BCS-GSM outperform the introduced reconstruction techniques, while Fig. 4.7 shows that we achieve better performance of 10% when using the Kalman filter.

4.6 Chapter Summary

In this chapter an information propagation method was introduced. First, we performed topic detection based on Joint Complexity and low dimensional classification based on Compressive Sensing with the accuracy of a Kalman filter as a refinement step. The experimental evaluation with a large set of tweets revealed a better performance, when compared with previous state-of-the-art methods, while being language-agnostic, without any need of grammar, dictionaries or semantics.

Chapter 5
Extension of Joint Complexity and Compressive Sensing

Abstract In this chapter, the theory of Joint Complexity and Compressive Sensing has been extended to three research subjects, (a) classification encryption via compressed permuted measurement matrices, (b) dynamic classification completeness based on Matrix Completion and (c) encryption based on the Eulerian circuits of original texts. In the first additional research subject we study the encryption property of Compressive Sensing in order to secure the classification process in Twitter without an extra cryptographic layer. The measurements obtained are considered to be weakly encrypted due to their acquisition process, which was verified by the experimental results. In the second additional research subject we study the application of Matrix Completion (MC) in topic detection and classification. Based on the spatial correlation of tweets and the spatial characteristics of the score matrices, we apply a novel framework which extends the Matrix Completion to build dynamically complete matrices from a small number of random sample Joint Complexity scores. In the third additional research subject, we present an encryption system based on Eulerian circuits, that destructs the semantics of a text while retaining it in correct syntax. We study the performance on Markov models, and perform experiments on real text.

5.1 Introduction

In this chapter, the theory of Joint Complexity and Compressive Sensing has been extended to three research subjects, (a) classification encryption via compressed permuted measurement matrices, (b) dynamic classification completeness based on Matrix Completion and (c) encryption based on the Eulerian circuits of original texts.

In the first additional research subject we study the encryption property of Compressive Sensing in order to secure the classification process in Twitter without an extra cryptographic layer. The measurements obtained are considered to be weakly encrypted due to their acquisition process, which was verified by the experimental results.

In the second additional research subject we study the application of Matrix Completion (MC) in topic detection and classification. Based on the spatial correlation of tweets and the spatial characteristics of the score matrices, we apply a novel

© Springer International Publishing AG 2018   69
D. Milioris, *Topic Detection and Classification in Social Networks*,
DOI 10.1007/978-3-319-66414-9_5

framework which extends the Matrix Completion to build dynamically complete matrices from a small number of random sample Joint Complexity scores.

In the third additional research subject, we present an encryption system based on Eulerian circuits, that destructs the semantics of a text while retaining it in correct syntax. We study the performance on Markov models, and perform experiments on real text.

5.2 Classification Encryption via Compressed Permuted Measurement Matrices

In this work we present an efficient encryption system based on Compressive Sensing for topic detection and classification. The proposed method first employs Joint Complexity to perform topic detection. Then based on the spatial nature of the data, we apply the theory of Compressive Sensing to perform classification from a small number of random sample measurements. The breakthrough of the method is the encryption based on the permutation of measurements which are generated when solving the classification optimization problem. The experimental evaluation with real data from Twitter presents the robustness of the encryption accuracy, without using a specific cryptographic layer, while maintaining a low computational complexity [63].

Similar to Chap. 4, we describe the theoretical concepts of the CS application [69, 71] in the problem of classification. Let $\mathbf{x} \in \mathbb{R}^N$ denote the signal of interest. Such signal can be represented as a linear combination of a set of basis $\{\psi_i\}_{i=1}^N$. By constructing an $N \times N$ basis matrix $\boldsymbol{\Psi} = [\psi_1, \psi_2, \dots, \psi_N]$, the signal x can be expressed as $x = \sum_{i=1}^N s_i \psi_i = \boldsymbol{\Psi} s$. In fact the signal is represented as $x = \boldsymbol{\Psi} s + \theta$, with $\theta \in \mathbb{R}^N$ being the noise, where $\mathbb{E}(\lesssim) = 0$ and $var(\theta) = O(|\boldsymbol{\Psi} s|)$.

The efficiency of a CS method for signal approximation or reconstruction depends highly on the sparsity structure of the signal in a suitable transform domain associated with an appropriate sparsifying basis $\boldsymbol{\Psi}$. The measurement model in the original space-domain is expressed as $\mathbf{g} = \boldsymbol{\Phi} \mathbf{x}$, where $\mathbf{g} \in \mathbb{R}^M$ is the measurement vector and $\boldsymbol{\Phi} \in \mathbb{R}^{M \times N}$ denotes the measurement matrix. The measurement model has the following equivalent transform-domain representation

$$\mathbf{g} = \boldsymbol{\Phi} \boldsymbol{\Psi} \mathbf{s} + \boldsymbol{\Phi} \theta \tag{5.1}$$

In fact when the length of the sequence $m \to \infty$ and $N \to \infty$, $\mathbb{E}(\boldsymbol{\Psi} s) = O(mN)$, with $var(\theta) = O(mN)$, $std(\theta) = O(\sqrt{|\Phi| m})$ and $\mathbb{E}(\boldsymbol{\Phi} \theta) = 0$. The second part of (5.1), $\boldsymbol{\Phi} \theta$ is of relative order $O\left(\frac{1}{\sqrt{mN}}\right)$, and is negligible compared to $\boldsymbol{\Phi} \boldsymbol{\Psi} s$ due to the law of large numbers.

Examples of measurement matrices $\boldsymbol{\Phi}$, which are incoherent with any fixed transform basis $\boldsymbol{\Psi}$ with high probability (universality property [18]), are random matrices with independent and identically distributed (i.i.d.) Gaussian or Bernoulli entries.

The problem of classifying a tweet is reduced to a problem of recovering the one-sparse vector $\mathbf{s}$. Of course in practice we do not expect an exact sparsity, thus, the estimated class corresponds simply to the largest-amplitude component of $\mathbf{s}$. According to [12, 18], $\mathbf{s}$ can be recovered perfectly with high probability by solving the following optimization problem

$$\hat{\mathbf{s}} = \arg\min_{\mathbf{s}} \Big(\|\mathbf{s}\|_1 + \tau \|\mathbf{g} - (\boldsymbol{\Phi}\boldsymbol{\Psi}\mathbf{s})\|_2 \Big) \tag{5.2}$$

where τ is a regularization factor that controls the trade-off between the achieved sparsity and the reconstruction error.

5.2.1 Preprocessing Phase

During the Preprocessing phase, we built our classes as described in Sect. 3.3 and for each class we extract the most representative tweet(s) (CTs) based on the Joint Complexity method. The vector $\boldsymbol{\Psi}_T^i$ consists of the highest JC scores of the i-th CT. The matrix $\boldsymbol{\Psi}_T$ is used as the appropriate sparsifying dictionary for the training phase. Moreover, a measurement matrix $\boldsymbol{\Phi}_T^i$ is associated with each transform matrix $\boldsymbol{\Psi}_T^i$, while T denotes the preprocessing phase.

The matrix $\boldsymbol{\Psi}_T^i \in \mathbb{R}^{N_i \times C}$ is used as the appropriate sparsifying dictionary for the i-th CT, since in the ideal case the vector of tweets at a given class j received from CT i should be closer to the corresponding vectors of its neighboring classes, and thus it could be expressed as a linear combination of a small subset of the columns of $\boldsymbol{\Psi}_T^i$. Moreover, a measurement matrix $\boldsymbol{\Phi}_T^i \in \mathbb{R}^{M_i \times N_i}$ is associated with each transform matrix $\boldsymbol{\Psi}_T^i$, where M_i is the number of CS measurements. In the proposed algorithm, a standard Gaussian measurement matrix is employed, with its columns being normalized to unit ℓ_2 norm. A random matrix or a PCA matrix could be also used.

5.2.2 Run Phase

A similar process is followed during the runtime phase. More specifically, we denote $\mathbf{x}_{c,R}$ as the Joint Complexity score of the incoming tweet with the CT_i classified at the current class c, where R denotes the runtime phase. The runtime CS measurement model is written as

$$\mathbf{g}_c = \boldsymbol{\Phi}_R \mathbf{x}_{c,R} \tag{5.3}$$

where $\boldsymbol{\Phi}_R^i \in \mathbb{R}^{M_{c,i} \times N_{c,i}}$ denotes the corresponding measurement matrix during the runtime phase. In order to overcome the problem of the difference in dimensionality between the preprocessing and run phase, while maintaining the robustness of the reconstruction procedure, we select $\boldsymbol{\Phi}_R^i$ to be a subset of $\boldsymbol{\Phi}_T^i$ with an appropriate number of rows such as to maintain equal measurement ratios.

The measurement vector $\mathbf{g}_{c,i}$ is formed for each CT i according to (5.3) and transmitted to the server, where the reconstruction takes place via the solution of (5.2), with the training matrix $\boldsymbol{\Psi}_T^i$ being used as the appropriate sparsifying dictionary. We emphasize at this point the significant conservation of the processing and bandwidth resources of the wireless device by computing only low-dimensional matrix-vector products to form $\mathbf{g}_{c,i}$ $(i = 1, \ldots, P)$ and then transmitting a highly reduced amount of data $(M_{c,i} \ll N_{c,i})$. Then, the CS reconstruction can be performed remotely (e.g. at a server) for each CT independently.

Last, we would like to note the assumption that the CS-based classification method involves the mobile device that collects the tweets from the Twitter API and a server that performs the core CS algorithm.

5.2.3   Security System Architecture

The method consists of two parts: (5.2.3.1) Privacy system, and (5.2.3.2) Key description.

5.2.3.1   Privacy System

Due to their acquisition process, CS measurements can be viewed as *weakly encrypted* for a malicious user without knowledge of the random matrices Φ^i, which have independent and identically distributed (i.i.d.) Gaussian or Bernoulli entries [18].

The encryption property of a CS approach relies on the fact that the matrix Φ is unknown to an unauthorized entity, since Φ can be generated using a (time-varying) cryptographic key that only the device and the server share.

More specifically, the server extracts the runtime sub-matrix Φ_R^i from the training Φ_T^i. The lines of Φ_R^i are permuted and the *key* of the merge of the false measurement vectors and the correct one is used, as shown in Figs. 5.1 and 5.2 and described extensively in [67, 71].

5.2.3.2   Key Description

The device sends the measurement vector g to the server along with $N - 1$ false vectors, where the reconstruction takes place. Then, the server uses the information

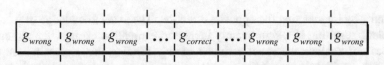

Fig. 5.1 $N - 1$ false vectors plus the correct one. This *key*, i.e., the sequence of the measurement vectors reaches the server

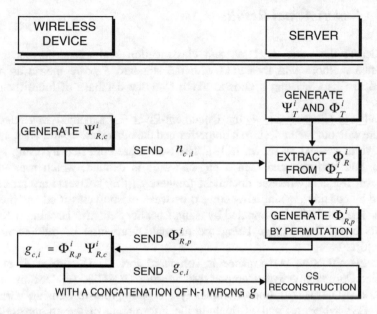

Fig. 5.2 System's security architecture

of the topic detection and the representative tweets based on their JC scores, etc., and performs classification as described in Sect. 4.3.

5.2.4 Possible Attacks from Malicious Users

A possible attack could follow two directions:

1. Find Φ matrix by intercepting the server. Modern network cryptographic protocols could guarantee that the decryption of Φ_p—where p denotes the permutation of the lines—is almost infeasible in practice due to the combinatorial nature of the inverse problem.

2. Find g by intercepting the opposite direction. The exact knowledge of g is insufficient, resulting in a significantly increased estimation error, when the attacker does not achieve the exact estimate of Φ_R^i. Finding the correct measurement vector g, which increases the estimation error (without an exact knowledge).

5.2.5   Experimental Results

Considering the topic detection and classification evaluation, the accuracy of the tested methods was measured with the standard *F-score* metric as already defined in Chap. 3, using a Ground Truth over the database of more than 1.5M tweets.

Similar to Chaps. 3 and 4, the Document-Pivot (DP) method was selected to compare with our method, since it outperformed the other state-of-the-art techniques in a Twitter context as shown in [2]. The tweets are collected by using specific queries and hashtags and then a bag-of-words is defined, which uses weights with term frequency-inverse document frequency (*tf-idf*). Tweets are ranked and merged by considering similarity context between existing classified and incoming tweets. The similarity is computed by using Locality Sensitive Hashing (LSH) [2], with its main disadvantage being the manual observation of training and test tweets [6].

The classification performance is compared for: (a) Document Pivot (DP), (b) Joint Complexity with Compressive Sensing (JC+CS), (c) Document Pivot with URL (DPurl), (d) Joint Complexity and Compressive Sensing with URL (JCurl+CS), where (c) and (d) include the information of the compressed URL of a tweet concatenated with the original tweet's text; extracted from the *.json* file.

Figure 5.3 compares the classification accuracy (increased by 37%) of the DP, DPurl and JC+CS, JCurl+CS method as a function of the number of measurements by using the ℓ_1-norm minimization. Figure 5.4 compares the reconstruction performance between several widely used norm-based techniques and Bayesian CS algorithms. More specifically, the following methods are employed[1]: (1) ℓ_1-norm minimization using the primal-dual interior point method (L1EQ-PD), (2) Orthogonal Matching Pursuit (OMP), (3) Stagewise Orthogonal Matching Pursuit (StOMP), (4) LASSO, (5) BCS [49], and (6) BCS-GSM [100, 101].

Figure 5.5 shows the encryption capability of the method for the Bayesian Compressive Sensing (BCS) [49] and Bayesian Compressive Sensing-Gaussian Scale Mixture (BCS-GSM) [100, 101] reconstruction algorithms, which outperformed in the classification accuracy.

---

[1]For the implementation of methods (1)–(5) the MATLAB codes can be found in: http://sparselab. stanford.edu/, http://www.acm.caltech.edu/l1magic, http://people.ee.duke.edu/~lcarin/BCS.html

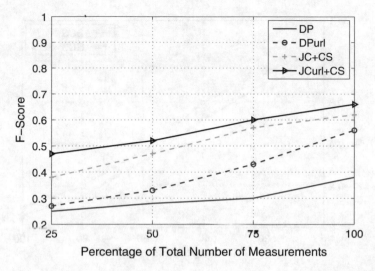

Fig. 5.3 Classification accuracy measured by F-Score for the DP, DPurl and JC+CS, JCurl+CS method as a function of the number of measurements (%) by using the ℓ_1-norm min

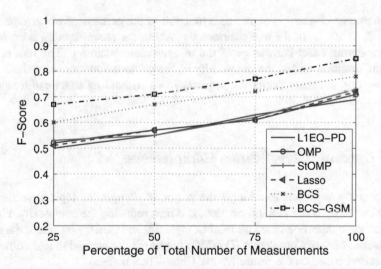

Fig. 5.4 Classification accuracy measured by F-Score as a function of the number of measurements (%) by using several reconstruction techniques, for the JCurl+CS method

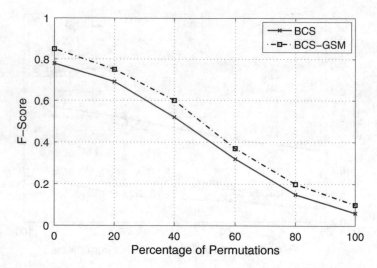

Fig. 5.5 Evaluation of the encryption property using BCS and BCS-GSM, for a varying number of permuted lines of Φ_R^i

The average *F-score* is shown as a function of the percentage of permuted lines from 0% to 100%, of the true matrices Φ_R^i, where the reconstruction is performed by considering exact knowledge of the measurement vectors g. The results agree with the intuition that as the complexity of the permutation increases, the *F-score* classification accuracy decreases (approx. 78%) without an exact estimate of the true measurement matrix.

5.3 Dynamic Classification Completeness

In this study we want to address the major challenges of topic detection and classification by using less information, and thus reducing the complexity. First we perform topic detection based on Joint Complexity and then we introduce the theory of dynamic Matrix Completion (DynMC) to reduce the computational complexity of JC scores based on the simple Matrix Completion method.

While there are recent works that propose solutions to the exhaustive computations [76] they are not taking into account the dynamics of the users and use synthetic data instead. We extend these methods by proposing a dynamic framework [51, 64, 70] that takes advantage of the spatial correlation of tweets, and thus reduces the computational complexity.

5.3.1 Motivation

The previously described procedure in Chaps. 3 and 4 requires the extensive calculation of JC scores, i.e. if there are N_s tweets in the s-th timeslot of the set of tweets to be classified, there are $N_s \times (N_s - 1)/2$ such scores to be calculated.

In order to address the exhaustive computation of JC scores, we perform random sampling on the score matrices. Random sampling reduces the time needed to build the score matrices and as a result the computation needed to check every tweet with every other tweet. In order to succeed on this, we need the existence of correlation between the matrix scores, which depend directly on the sequences (tweets) we have in our database's timeslots.

Tweets have spatial correlation since the ones closer to the meaning show similar measurement score matrices. Since the tweets are correlated, the degrees of freedom of the score matrices are much lower than their dimension. If a matrix has a low rank property, then it presents a limited number of degrees of freedom.

While the recovery of the $i \cdot j$ entries of matrices $s_1, s_2, \ldots, s_S$ is impossible from a number of measurements m (where $m \ll i \cdot j$), MC [11] shows that such a recovery is possible when the rank of a matrix s is small enough compared to its dimensions. In fact, the recovery of the unknown matrix is feasible from $m \geq cj^{6/5}r\log j$ random measurements, where $j > i$ and $\text{rank}(s) = r$. The original matrix s can be recovered by solving the following optimization problem:

$$\min ||s||_* \text{ s.t. } A(s) = A(M) \tag{5.4}$$

where $||s||_* = \sum_{k=1}^{\min\{i,j\}} \sigma_k(s)$ with $\sigma_k(s)$ being the k-th largest singular value of s. M is the matrix s after sub-sampling, while A is a linear map from $R^{i \cdot j} \rightarrow R^m$, that has uniform samples in rows and columns and satisfies the Restricted Isometry Property.

Sampling on the measurements vectors of s will provide an incomplete scores matrix. The topic that has to be revealed uses a subset Ω of the measurements of S, which randomly chooses by sensing a random number of the $k < h$ tweets of the timeslot. Finally the topic detection get $\Omega \subseteq |i| \times |j|$ measurements, with

$$|\Omega| = \frac{k(i \times j)}{h} \tag{5.5}$$

while the sampling map $A_\Omega(M)$ has zero entries at the j-th position of the i-th timeslot if $s(i, j) \notin \Omega$.

During the runtime phase we need to recover the unobserved measurements of matrix s, denoted by s^-, by solving the following minimization problem

$$\min ||s^-||_* \text{ s.t. } ||A_\Omega(s^-) - A_\Omega(M)||_F^2 < \epsilon \tag{5.6}$$

where F denotes the Euclidean norm, and ϵ is the noise parameter. The convex optimization problem in (5.6) can be solved by an interior point solver, e.g. CVX [28], or via singular value thresholding, e.g. FPC and SVT [10], which applies a singular value decomposition algorithm and then projection on the already known measurements in each step.

5.3.2 Proposed Framework

In this section, we describe our proposed framework, DynMC, led by the intuition of the spatio-temporal correlations between JC scores among the several representative tweets. During the training phase we collect tweets and compute the JC scores at each time t.

Assume that $C \in R^{i \times i}$ defines the temporal correlation of the tweet in specific classes, while ϵ indicates the noise. The relationship of the tweets between the JC scores and the representative tweets over time can be expressed as:

$$[A(M)]^t = C\,[A(M)]^{t-1} + \epsilon \tag{5.7}$$

where $[A(M)]^t$ and $[A(M)]^{t-1} \in R^{i \times 1}$ represent the JC scores at time t and $t-1$, respectively, received at a specific class.

As it was mentioned earlier, tweets have a spatial correlation, since closer tweets or classes show similar measurement vectors. In this study we try to address this problem by introducing a dynamic Matrix Completion technique. The proposed technique is able to recover the unknown matrix at time t by following a random sampling process and reduce the exhaustive computation of JC scores.

As it was mentioned in Sect. 5.3.1 subsampling gives matrix M_t at each time t of the sampling period and we receive a subset $\Omega_t \subseteq |i| \times |j|$ of the entries of M_t, where $|\Omega|_t = k \times i$. The sampling operator A_t (as defined in Sect. 5.3.1) gives

$$[A_t(M)]_{j,i} = \left\{ P_{j,i}, (j, i) \in \Omega_t 0, \text{otherwise.} \right. \tag{5.8}$$

where $P_{j,i}$ is the JC score received at j, i and Ω_t is a subset of the complete set of entries $|i| \times |j|$, where $\Omega_t \cup \Omega_t^C = |i| \times |j|$.

While the sampling operator $A_t^C(M_t)$ collects the unobserved measurements at time t, we also define the sampling operator $A_I = A_{t-1} \cap A_t^C$ as the intersection of the training measurements of the classes by time $t-1$.

We need to recover the fingerprint map M_t that will be used during the runtime phase by taking into account the JC scores received on previous time windows. The proposed technique reconstructs matrix M_t that has the minimum nuclear norm, subject to the values of $M_t \in \Omega_t$ and the sampled values at time $t-1$. There is a clear correlation with measurements at time t via C according to the model in Eq. (5.7).

Matrix C and the original matrix M_t can be recovered by solving the following optimization problem

$$\min_{\tilde{M}_t, C} ||\tilde{M}_t||_* \text{ s.t.} \tag{5.9}$$

$$||A_t(\tilde{M}_t) - A_t(M_t)||_F^2 \le \epsilon_1 \, ||A_I(C \cdot M_{t-1}) - A_I(M_t)||_F^2 \le \epsilon_2 \tag{5.10}$$

where $\tilde{M}_t$ is the recovered JC scores matrix at time t. Variables $\epsilon_1, \epsilon_2 \ge 0$ represent the tolerance in approximation error, while $|| \cdot ||_F$ denotes the Frobenious norm as mentioned in Sect. 5.3.1. Matrix C expresses the relationship between the values of $M_t \in \Omega_t^C \cap \Omega_{t-1}$ and it is adjusted to the number of common tweets at time $t - 1$ and t. CVX [28] can be used to solve the general convex optimization problem in Eqs. (5.10) and (5.10).

5.3.3   Experimental Results

Considering the topic detection and classification evaluation, the accuracy of the tested methods was measured with the standard *F-score* metric, using a ground truth over the database of more than 1.5M tweets. The Document-Pivot method was selected to compare with our method, since it outperformed the other state-of-the-art techniques in a Twitter context as shown in [2]. The tweets are collected by using specific queries and hashtags and then a bag-of-words is defined, which uses weights with term frequency-inverse document frequency (*tf-idf*). Tweets are ranked and merged by considering similarity context between existing classified and incoming tweets. The similarity is computed by using Locality Sensitive Hashing (LSH) [2], with its main disadvantage being the manual observation of training and test tweets [6].

The classification performance is compared for: (a) Document Pivot (DP), (b) Joint Complexity with Compressive Sensing (JC+CS) [12, 69], (c) Document Pivot with URL (DPurl), (d) Joint Complexity and Compressive Sensing with URL (JCurl+CS) [12, 69], where (c) and (d) include the information of the compressed URL of a tweet concatenated with the original tweet's text; extracted from the *.json* file.

Figure 5.6 shows the recovery error of the score matrices s based on FPC algorithm. We can recover the s_i matrices by using approximately 77% of the original symmetric part with the error of *completeness* $\to 0$, while addressing the problem of exhaustive computations.

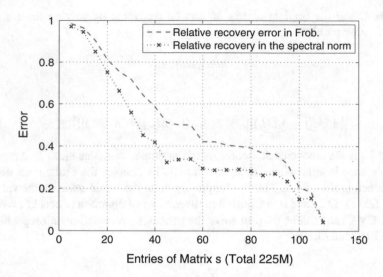

Fig. 5.6 Reconstruction error of s_i by using the FPC algorithm. Approximately 77% of the symmetric part needed while the error of *completeness* $\rightarrow 0$

Several widely used norm-based techniques and Bayesian CS algorithms are employed to treat the classification problem[2]: (1) ℓ_1-norm minimization using the primal-dual interior point method (L1EQ-PD), (2) Orthogonal Matching Pursuit (OMP), (3) Stagewise Orthogonal Matching Pursuit (StOMP), (4) LASSO, (5) BCS [49] and (6) BCS-GSM [100, 101].

Figure 5.7 compares the topic detection accuracy decreased by 10% for the DP and DPurl and 5% for the JC and JCurl method as a function of the number of measurements of the reconstructed matrix s^- (67% of **s**) by using the ℓ_1-norm min. JC and JCurl uses the Bayesian framework which makes them more robust to the noisy process of scores matrices computation while exploring the spatial correlation of the measurements.

Finally, Fig. 5.8 shows the performance of the proposed DynMC method versus the classic MC as a function of the total number of measurements of the recon-structed symmetric matrix. We can observe a faster convergence of the DynMC method as the number of measurements is increased. More specifically, after the critical point of 40% of the symmetric part, the DynMC performs better and achieves completeness at 78% of the original matrix.

---

[2]For the implementation of methods (1)–(5) the MATLAB codes can be found in: http://sparselab. stanford.edu/, http://www.acm.caltech.edu/l1magic, http://people.ee.duke.edu/~lcarin/BCS.html

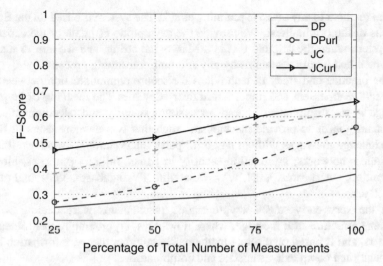

Fig. 5.7 Topic detection accuracy measured by F-Score for the DP, DPurl and JC, JCurl method as a function of the number of measurements (%) on the recovered matrix s^- by using the ℓ_1-norm min. on 67% of the measurements

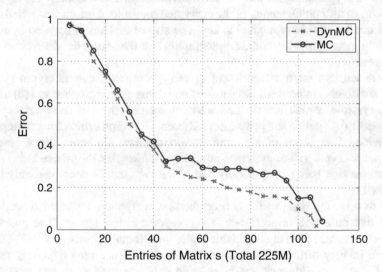

Fig. 5.8 Reconstruction error of s_i by using the FPC algorithm. DynMC has a faster convergence compared to MC as the error of *completeness* $\rightarrow 0$

5.4 Stealth Encryption Based on Eulerian Circuits

In this work, we introduce an encryption system that destructs the semantics of a text while retaining it almost in correct syntax. The encryption is almost undetectable, since the text is not able to be identified as different to a regular text. This makes the system resilient to any massive scanning attack. The system is based on the Eulerian circuits of the original text. We provide the asymptotic estimate of the capacity of the system when the original text is a Markovian string, and we aim to make the encrypted text hardly detectable by any automated scanning process.

The practice and study of techniques for secure communication between users has become emergent and important in every day life. The meaning of cryptology is huge in the disciplines of computer systems security and telecommunications. The main goal is to provide mechanisms such that two or more users or devices can exchange messages without any third party intervention. Nowadays, it is used in wireless networks, information security in banks, military purposes, biometric recognition, smart cards, VPN, WWW, satellite TV, databases, VOIP and plethora of systems.

In the very early stages, cryptographic mechanisms were dealing with the language structure of a message, where nowadays, cryptography has to deal with numbers, and is based on discrete mathematics, number theory, information theory, computational complexity, statistics and combinatorics.

Cryptography consists of four basic functions: (a) confidentiality, (b) integrity, (c) non-repudiation and (d) certification. The encryption and decryption of a message is based on a cryptographic algorithm and a cryptographic key. Usually the algorithm is known, so the confidentiality of the encrypted transmitted message is based on the confidentiality of the cryptographic key. The size of that key is counted in number of bits. In general, the larger the cryptographic key, the harder the decryption of the message.

There are two main categories of crypto-systems: (a) classic crypto-systems, which are divided into substitution ciphers and transposition ciphers, and (b) modern crypto-systems, which are divided into symmetric (share a common key) and asymmetric (use public and private key). Systems based on symmetric cryptography are divided into block ciphers and stream ciphers. However, users based on asymmetric cryptographic systems know the public key, but the private key is secret. The information being encrypted by one of the keys can be decrypted only by the other key.

Up to now, the main methods used for text encryption are sophisticated algorithms that transform original data into encrypted binary stream. The problem is that such streams are easily detectable under an automated massive attack, because they are in very different format and aspect of non-encrypted data, e.g. texts in natural language. This way any large scale data interception system would very easily detect the encrypted texts. The result of this detection is twofold. First detected encrypted texts can be submitted to massive decryption processes on large computing resources, and finally be deciphered. Second, even when the texts would not eventually be deciphered, the source or the destination of the texts are at least identified as hiding their communications and therefore can be subject to other intrusive investigations. In other words, encryption does not protect against a massive spying attack if encrypted texts are easily detectable.

In our method we use a permutation of the symbols i.e. n-grams of the original text. Doing so leads to the apparent destruction of the semantic information of the text while keeping the text quasi correct in its syntax, and therefore undetectable under an automated syntactic interception process. To retrieve the original information the n-grams would be reordered in their original arrangement.

5.4.1   Background

In the following terminology, let T be a text written on an alphabet $\mathscr{A}$ of size V. Let r be an integer, which is used to denote a sequence of r consecutive symbols appearing in a text, i.e. r-gram, of text T.

5.4.1.1   Syntax Graph

According to the terminology given earlier, we define the *syntax graph* G of text T. We assume a fixed integer r and we denote $G_r(T) = (V, E)$ the directed *syntax graph* of the r-grams of text T. There is an edge between two r-grams a and b, if b is obtained by the translation by one symbol of r-gram a in the text T. For example, the 3-gram $b =$ "*mpl*" follows the 3-gram $a =$ "*amp*" in the text $T =$ "*example*".

The graph $G_r(T)$ is a multi-graph, since several edges can exist between two r grams a and b, as many as b follows an instance of a in the text T.

Figure 5.9 shows the syntax graph of the famous Shakespeare's text T = "*to be or not to be that is the question*", with $|V| = 13$ (for 1-grams) and $|E| = 39$.

5.4.1.2   Eulerian Path and Circuit

An Eulerian path in a multi-graph is a path which visits every edge exactly once. An Eulerian circuit or cycle is an Eulerian path which starts and ends on the same vertex and visits every edge exactly once. The number of Eulerian paths in a given multi-graph is easy to compute (in a polynomial time) as a sequence of binomials and a determinant; or can be adapted from BEST theorem [1, 99], to enumerate the number of Eulerian circuits [35, 42, 60] which will be explained in Sect. 5.4.3.

5.4.2   Motivation and Algorithm Description

5.4.2.1   Motivation

The originality of the system is that the appearance of the transmitted information, although encoded, is not changed, so that malicious software applications cannot determine that a flow of information under study is encoded or not. Consequently, being undetected, the encoded information will not be subject to massive decipher-

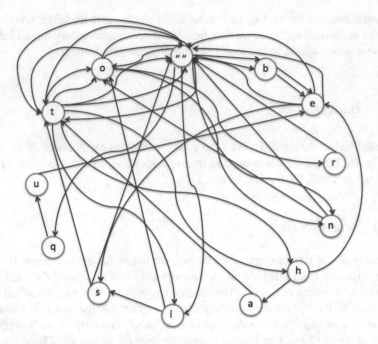

Fig. 5.9 Syntax Graph (1-grams) of the famous Shakespeare's sentence T = *"to be or not to be that is the question"*, where *" "* denotes the space

ing techniques. The probability of the transmitted information being deciphered is therefore dramatically higher, regardless of the encoding scheme.

When a malicious deciphering application has detected an emitter or a receiver as transmitting encoded information (regardless of whether or not it managed at deciphering it), it may tag it, so as to trigger further actions, like systematically analysing of its transmitted information, trying to infiltrate it to steal further information. Therefore, an additional and un-addressed technical problem solved by this system consists in avoiding being determined by such an application the fact that encoded information is transmitted.

The system applies in particular to text transmission, but can also apply to other types of information. The encoded text comprises of only a few grammatical or lexicographic errors, so that any detecting mechanisms based on first level errors will not detect anything and will consider such an encoded text as "badly written normal text". Accordingly, the encoded text will not be detected as encoded and no deciphering mechanism will be triggered. Any malicious application analysing such information will detect a text, with a typical text format (words, sentences, etc.), and detecting that it is encoded text would require a complex semantic analysis. In addition, such a semantic analysis would become more difficult by having no information about the language used to write such a text.

Let the text "*Up to now, the main ... easily detectable*". written in the Introduction, be the text T. We design the Syntax Graph $G(T)$ with the 4-grams of T, which gives 3.42×10^{42} discrete Eulerian circuits. One possible circuit is the following:

T' = "In In other words encryptions and finally be deciphered, therefore can be submitted binary stream. This way any large communication does not protectable under an automated massive investination is twofold. First detect the texts can be decrypted texts are in very easily detection of this detectable. Up to now, they are easily detected encryption processes on large scale data into encrypted to massive deciphered. Second, even when the main methods used form original data intrusive spying resources, and as hiding attack if encrypted data, e.g. text encrypted aspect of non-encrypted texts are attack, because the source or texts would very different for the encryption are easily detect against a massive at least identified algorithms that transformat and their computing the texts in natural language. The problem is that such streams are sophisticated texts. The result of the destigation system would not eventually be subject to other interceptions. In"

In this example, it is clear that the encoded text looks like an English text, and that it would require either manual intervention to determine that it is not a normal text, or a very complex automatic semantic analysis. Even the small text presented in Fig. 5.9 gives 1.19×10^{10} discrete Eulerian circuits.

5.4.2.2   Algorithm Description

In general, the block ciphers method, breaks into blocks the initial message, which is going to be encrypted, and encrypts every block separately. Usually these blocks's size is 64 or 128 bits. This way the encryption can be adapted to the length of the information to be coded. The encryption is based on a mathematical encryption or cryptographic function $f(x, k_0)$ where x is the block and k_0 is a shared key. The result is a new cryptographic block y that usually has the same length with the initial block.

The length of every block has to be large enough, in order to avoid dictionary attacks. If the length of every block is small, then a malicious user can find a pair of clean and respectfully encrypted block to design a dictionary that corresponds every clean block to a encrypted block. Based on that dictionary, every text that is encrypted with the particular key could be decrypted.

The decryption process is the inverted encryption process $g(y, k_0)$ such that $g(f(x, k_0), k_0) = x$. In that case, a decryption function is used instead of the encryption function. Some classic block cipher algorithms are Data Encryption Standard [73], Triple DES which uses three keys instead of one, Advanced Encryption Standard [16], Tiny Encryption Algorithm [104] and others [79].

In our algorithm we assume that there is an encryption function $f(x, N, k_0)$ where k_0 is a shared key that produces an integer $y \in (0, N-1)$, when x is an integer, and $\in (0, N-1)$. Let $g(y, N, k_0)$ be an inverse function, such that $x = g(f(x, N, k_0), N, k_0)$. The function can be made of the concatenation of m blocks encryption of size ℓ (64 or 128 bits) with $m = \lceil \frac{\log_2 N}{\ell} \rceil$. In this case $f(x, N, k_0) = x_1 y_2 \ldots y_m$ if x is made of the concatenation of blocks $x_1 \ldots x_m$ with $y_i = f(x_i, k_0)$ for all $1 < i < m$. The first block is left unencrypted in order to have $y < N$ with very high probability.

Let r_0 be an integer, the proposed scheme is described in the following algorithm:

Algorithm 4 Encryption system based on Eulerian path

Suppose that Alice is the transmitter with text T, Bob is the receiver, and Neysa is the malicious user.

1. Transmitting Information:

    - Alice builds the syntax graph $G_{r_0}(T)$
    - Computes the number N of Eulerian paths. *Assume that Alice and Bob share the same indexing method of the Eulerian paths in $G_{r_0}(T)$.*
    - *Since text T itself is a specific Eulerian path, assume that $I(T)$ is the index of text T as Eulerian path.* Alice builds the text T' being the Eulerian path with index $f(I(T), N, k_0)$, and then transmits the text T' to Bob.

    According to the malicious user Neysa, the text T' seems to have correct syntax, but the original semantic is destroyed.

2. Receiving Information:

    - Bob builds the syntax graph $G_{r_0}(T')$ which is the same as $G_{r_0}(T)$.
    - Computes the number of Eulerian paths N and is able to recover the original text index $I(T) = g(I(T'), N, k_0)$.

5.4.3 Performance in Markov Models

In this section we consider a string $T = X_n$ of length n generated by a Markov process of memory 1 over an alphabet $\mathscr{A}$ of size V. Notice that there is no loss of generality with the r-grams description of our model, in this case $\mathscr{A}$ is the set of all r-grams. We denote $\mathbf{P}$ the Markov transition matrix: for $(a, b) \in \mathscr{A}^2$ we denote p_{ab} the coefficients of matrix $\mathbf{P}$, i.e. the probability that symbol b follows symbol a.

The Euler code length of a string is the logarithm in base 2 of the number of eulerian circuits that can be built from string X_n while making it cyclic. We denote this quantity L_n it reflects the number of bits needed to index the eulerian number and indeed the encoding capacity of our scheme.

Theorem 5.1 *The average code length L_n of X_n is asymptotically equivalent to the entropy $H(X_n)$ of the string X_n when $n \to \infty$.*

This theorem has an important consequence. Let denote Y_n the string obtained by the encryption algorithm. We assume that the encryption algorithm selects an Euler circuit uniformly in the set of Eulerian circuits that can be built in X_n, then we have the following corollary:

Corollary 5.1 *The mutual information rate $\frac{1}{n}I(X_n, Y_n)$ tends to zero when $n \to \infty$.*

Proof We already know that $H(Y_n) = H(X_n)$ since X_n and Y_n have same probability in the Markov process, thus $I(X_n, Y_n) = H(X_n) - H(Y_n|X_n)$, where the last term is the conditional entropy of Y_n with respect to X_n. Since $H(Y_n|X_n) = L_n$ the results hold.

Remark

The order of magnitude of $I(X_n, Y_n)$ that is obtained in the proof of Theorem 5.1 is $O(\log^3 n\sqrt{n})$ but the right order should be $O(\log n)$ with a more careful handling of the error terms. Anyhow this proves that our scheme is efficiently erasing the information of X_n while keeping the text in a formal shape.

We will also evaluate the average number E_n of Eulerian circuits in X_n. To this end, for $s \geq 0$ we denote $\mathbf{P}(s)$ the matrix made of coefficients $(p_{ab})^s$ (if $p_{ab} = 0$ we assume $(p_{ab})^s = 0$). We denote $\lambda(s)$ the main eigenvalues of matrix $\mathbf{P}(s)$.

Theorem 5.2 *The average number of Eulerian circuits of string X_n of length n is equivalent to $\frac{\alpha}{n}\lambda^{2n}(\frac{1}{2})$ for some $\alpha > 0$ that can be explicitly computed.*

Remark

This average is purely theoretical, since it is impossible to simulate this result when n is large like in the example text, since the most important and decisive contributions come from strings with extremely low probabilities. In order to prove our theorem we need some notations and lemmas.

Let $\mathbf{k}$ be a $V \times V$ integer matrix defined on $\mathscr{A} \times \mathscr{A}$ which is the adjacency matrix of syntax graph $G(X_n)$, i.e. the coefficient k_{ab} of $\mathbf{k}$ is equal to the number of time symbol b follows symbol a in X_n, we say that $\mathbf{k}$ is the *type* of string X_n as defined in [42]. For $(a, b) \in \mathscr{A}^2$ we also denote δ_{ab} the type of the string ab.

For $c \in \mathscr{A}$ we denote $k_c = \sum_{d \in \mathscr{A}} k_{cd}$ and $k^c = \sum_{d \in \mathscr{A}} k_{dc}$ respectively the outdegree and indegree of symbol c in the syntaxic graph. Let $\mathscr{F}_n$ the set of *balanced* types, i.e. such that $\forall c \in \mathscr{A} : k_c = k^c$, and such that $\sum_{(c,d) \in \mathscr{A}^2} k_{cd} = n$.

Lemma 5.1 *The set $\mathscr{F}_n$ is a lattice of dimension $V^2 - V$ [42]. Its size $a(n) = O(n^{V^2-V+1})$ and we denote ω the volume of its elementary element.*

Proof The set of matrix is embedded in the vector space of real matrices which is a dimension V^2. The V balance equations are in fact $V - 1$ since any of them can be deduced from the sum of the other. There is a last equation to specify that all coefficients sum to n.

We denote $\mathscr{F}(1)$ the set of balanced real matrices with positive coefficients that sum to 1. For $\mathbf{y}$ a real non-negative matrix and $s \geq 0$, we denote $L(\mathbf{y}, s) = \sum_{(c,d) \in \mathscr{A}^2} y_{c,d} \log \left(\frac{y_c}{y_{cd}} p_{cd}^s \right)$

We have the following technical lemma.

Lemma 5.2 *Let $s > 0$, the maximal value of $L(\mathbf{y}, s)$ for $\mathbf{y} \in \mathcal{F}(1) - \frac{\delta_{ba}}{n}$ is the matrix $\tilde{\mathbf{y}}_n(s)$ which converges to a matrix $\tilde{\mathbf{y}}(s) \in \mathcal{F}(1)$ whose (c, d) coefficient is $v_c(s) u_d(s) \frac{p_{cd}^s}{\lambda(s)}$ with $(u_c(s))_{c \in \mathcal{A}}$ and $(v_c(s))_{c \in \mathcal{A}}$ being respectively the right and left main eigenvectors of $\mathbf{P}(s)$ (with $\sum_{c \in \mathcal{A}} u_c(s) v_c(s) = 1$). For instance, $L(\tilde{\mathbf{y}}(s), s) = \log \lambda(s)$ and $L(\tilde{\mathbf{y}}_n(s), s) = L(\tilde{\mathbf{y}}(s), s) - \frac{1}{n} \left(\log \lambda(s) \frac{u_b(s)}{u_a(s)} \right)$.*

Proof We have for all $(c, d) \in \mathcal{A}^2$ the gradient matrix is

$$\frac{\partial}{\partial y_{cd}} L(\mathbf{y}, s) = \log \frac{y_c}{y_{cd}} p_{cd}^s . \tag{5.11}$$

The maximum on $\mathcal{F}(1)$ or $\mathcal{F}(1) - \frac{1}{n} \delta_{ba}$ must be member of the vector space generated by the matrix $\mathbf{1}$ (made of all one), and the matrices $\mathbf{A}_j, j \in \mathcal{A}$, the coefficients of A_j are all zeros excepted 1 on the jth column and -1 on the jth row, and zero on the diagonal. These matrices are the orthogonal matrices that define $\mathcal{F}(1)$ (or a translation of it). Membership to this vector space is equivalent to the fact that $\frac{\partial}{\partial y_{cd}} L(\mathbf{y}, s)$ must be of the form $\alpha + z_c - z_d$ for some α and $(z_c)_{c \in \mathcal{A}}$, which is equivalent to the fact that

$$\frac{y_{cd}}{y_c} = \frac{x_d}{x_c} \frac{p_{cd}^s}{\lambda} \tag{5.12}$$

for some λ and $(x_c)_{c \in \mathcal{A}}$. From the fact that $\sum_{d \in \mathcal{A}} \frac{y_{cd}}{y_c} = 1$ we get $\lambda = \lambda(s)$ and $(x_c)_{c \in \mathcal{A}} = (u_c(s))_{c \in \mathcal{A}}$, i.e. $\lambda x_c = \sum_{d \in \mathcal{A}} p_{cd}^s x_d$. Consequently

$$L(\mathbf{y}, s) = \sum_{(c,d) \in \mathcal{A}^2} y_{cd} \log(\lambda \frac{x_c}{x_d}) \tag{5.13}$$

$$= \log(\lambda) \sum_{(c,d) \in \mathcal{A}^2} y_{cd} \tag{5.14}$$

$$+ \sum_{c \in \mathcal{A}} (y_c - y_c) \log(x_c) \tag{5.15}$$

Thus $L(\tilde{\mathbf{y}}(s), s) = \log \lambda(s)$ and $L(\tilde{\mathbf{y}}_n(s), s) = (1 - \frac{1}{n}) \log \lambda(s) + \frac{1}{n} \log \frac{u_a(s)}{u_b(s)}$.

To simplify our proofs we will assume in the sequel that all strings start with a fixed initial symbol a. The strings starting with a having type $\mathbf{k}$ have probability $\prod_{(c,d) \in \mathcal{A}^2} p_{cd}^{k_{cd}}$ that we denote $\mathbf{P}^\mathbf{k}$. We denote

$$B_\mathbf{k} = \prod_{c \in \mathcal{A}} \binom{k_c}{(k_{cd})_{b \in \mathcal{A}}} . \tag{5.16}$$

For a matrix $\mathbf{M}$ and $(c, d) \in \mathcal{A}^2$ we denote $\det(\mathbf{M})$ the (c, d) cofactor of $\mathbf{M}$. We know that the number of eulerian circuits that share the same balanced type $\mathbf{k}$ is equal to $B_{\mathbf{k}} \frac{\det_{cd}(\mathbf{I}-\mathbf{k}^*)}{k_c}$ for any pair of symbols $(c, d) \in \mathcal{A}^2$, defining $\mathbf{k}^*$ as the matrix $\frac{k_{cd}}{k_c}$ for $(c, d) \in \mathcal{A}^2$ and $\mathbf{I}$ the identity matrix.

The number of strings that share the same type $\mathbf{k}$ depends on their terminal symbol b. Let denote $N_{\mathbf{k}}^b$ this number. When $b \neq a$, the type is not balanced but $\mathbf{k} + \delta_{ba} \in \mathcal{F}_n$. As described in [42],

$$N_{\mathbf{k}}^b = B_k \det_{bb}(\mathbf{I} - (\mathbf{k} + \delta_{ba})^*) . \tag{5.17}$$

In passing $N_{\mathbf{k}}^b \mathbf{P}^{\mathbf{k}}$ is the probability that string X_n has type $\mathbf{k}$ and we have the identity

$$\sum_{b \in \mathcal{A}} \sum_{\mathbf{k} + \delta_{ba} \in \mathcal{F}_n} N_{\mathbf{k}}^b \mathbf{P}^{\mathbf{k}} = 1 . \tag{5.18}$$

Similarly the number of eulerian circuits $E_{\mathbf{k}}^b$ corresponding to the string X_n assuming it ends with b and is made cyclic satisfies:

$$E_{\mathbf{k}}^b = \frac{1}{k_{ba} + 1} B_k \det_{bb}(\mathbf{I} - (\mathbf{k} + \delta_{ba})^*) . \tag{5.19}$$

For convenience we will handle only natural logarithms. We have $L_n = \sum_{b \in \mathcal{A}} L_n^b$ with

$$L_n^b = \sum_{\mathbf{k} + \delta_{ba} \in \mathcal{F}_n} N_{\mathbf{k}}^b \mathbf{P}^{\mathbf{k}} \log E_{\mathbf{k}}^b . \tag{5.20}$$

Proof (Proof of Theorem 5.1) Using the Stirling approximation: $k! = \sqrt{2\pi k} k^k e^{-k}(1 + O(\frac{1}{k}))$ and defining $\ell(\mathbf{y}) = \sum_{cd} y_{cd} \log \frac{y_c}{y_{cd}}$ we have

$$L_n^b = (n + O(1)) \frac{1}{n^{(V-1)V/2}} \sum_{\mathbf{k} + \delta_{ba} \in \mathcal{F}_n} r_b(\mathbf{y}) \exp(nL(\mathbf{y}, 1)) \ell(\mathbf{y})$$

$$+ O(\log n) \tag{5.21}$$

where $r_b()$ is a rational function and $\mathbf{y}$ is the matrix of coefficients $y_{cd} = \frac{k_{cd}}{n}$.

According to Lemma 5.2 the maximum value of $L(\mathbf{y}, 1)$ is $\log \lambda(1) = 0$ and thus we already have $L_n^b = O(a(n)) = O(n^{V^2 - V})$. For a matrix $\mathbf{M}$ we denote $\|\mathbf{M}\|$ the cartesian norm, i.e. the square root of the squared coefficients. Since $\tilde{\mathbf{y}}(1)$ is the maximum of $L(\mathbf{y}, 1)$ over $\mathcal{F}(1)$ there exists $A > 0$ such that $\forall \mathbf{y} \in \mathcal{F}(1)$: $L(\mathbf{y}, 1) \leq L(\tilde{\mathbf{y}}(1), 1) - A\|\mathbf{y} - \tilde{\mathbf{y}}(1)\|^2$, and when $\mathbf{y} + \frac{1}{n} \delta_{ba} \in \mathcal{F}(1)$

$$L(\mathbf{y}, 1) \leq L(\tilde{\mathbf{y}}_n(1)) - A\|\mathbf{y} - \tilde{\mathbf{y}}_n(1)\|^2 \tag{5.22}$$

Let $B > 0$: we define

$$L_n^b(B) = \frac{n}{n^{(V-1)V/2}} \sum_{\substack{k+\delta_{ba}\in\mathscr{F}_n \\ |\mathbf{y}-\tilde{\mathbf{y}}_n|<B\log n/\sqrt{n}}} r_b(\mathbf{y})\exp(nL(\mathbf{y}))\ell(\mathbf{y}) \tag{5.23}$$

From (5.22) we have $L_n^b = L_n^b(B)(1 + O(n^{-1}))$. Since function $L(\mathbf{y}, 1)$ is infinitely derivable and attains its maximum in $\mathscr{F}(1)$, which is zero, on $\tilde{\mathbf{y}}_n(1)$ we have for all $\mathbf{y} \in \mathscr{F}(1) - \frac{1}{n}\delta_{ba}$

$$L(\mathbf{y}, 1) = L(\tilde{\mathbf{y}}_n(1)) - D_n(\mathbf{y} - \tilde{\mathbf{y}}_n(1)) + O(\|\mathbf{y} - \tilde{\mathbf{y}}_n(1)\|^3), \tag{5.24}$$

where $D_n()$ is a positive quadratic form obtained from the second derivative of $L(\mathbf{y}, 1)$ on $\tilde{\mathbf{y}}_n(1)$. The value of L_n will be attained in the vicinity of the maximum since $\exp(nL(\mathbf{y}, 1))$ behaves like a Dirac when $n \to \infty$. Indeed we are in a kind of Saddle point application. Thus, since $L(\tilde{\mathbf{y}}_n(1)) = \frac{1}{n}\log\lambda(1)\frac{u_b(1)}{u_a(1)} = 0$ ($\lambda(1) = 1$ and $\forall c \in \mathscr{A}: u_a(1) = 1$)

$$L_n^b(B) = \frac{1}{n^{(V-1)V/2}}\left(1 + O\left(\frac{\log^3 n}{\sqrt{n}}\right)\right)$$

$$\sum_{\substack{k+\delta_{ba}\in\mathscr{F}_n \\ |\mathbf{y}-\tilde{\mathbf{y}}_n|<B\log n/\sqrt{n}}} \tilde{r}(\mathbf{y})e^{nD_n(\mathbf{y}-\tilde{\mathbf{y}}_n(1))}. \tag{5.25}$$

Since $\frac{1}{n}\mathscr{F}_n$ is a lattice of degree $(V-1)V$ with elementary volume ωn^{-V^2+V} and $D_n()$ converge to some the non-negative quadratic form $D()$ on $\mathscr{F}(1)$:

$$L_n^b(B) = n\ell(\tilde{\mathbf{y}}(1))\frac{r(\tilde{\mathbf{y}}(1))}{n^{(V-1)V/2}}\frac{n^{V^2-V}}{\omega}\left(1 + O\left(\frac{1}{\sqrt{n}}\right)\right)$$

$$\int_{\mathscr{F}(1)} e^{-nD(\mathbf{y}-\tilde{\mathbf{y}}(1))}d\mathbf{y}^{V(V-1)} \tag{5.26}$$

We use the property that

$$\int_{\mathscr{F}(1)} \exp(-nD(\mathbf{y}-\tilde{\mathbf{y}}))d\mathbf{y}^{V(V-1)} = \frac{1}{\sqrt{n^{V^2-V}\det(\pi D)}} \tag{5.27}$$

where $\det(\pi D)$ must be understood as the determinant of the quadratic operator $\pi D()$ on the vector space $\mathscr{F}(1)$. Therefore

$$L_n = n\ell(\mathbf{y})\frac{1}{\sqrt{\det(\pi D)}}\sum_{b\in\mathscr{A}} r_b(\tilde{\mathbf{y}})\left(1 + O\left(\frac{\log^3 n}{\sqrt{n}}\right)\right). \tag{5.28}$$

The same analysis can be done by removing $\log E_{\mathbf{k}}^b$ and since via (5.18) we shall get $\sum_{b \in \mathscr{A}} \frac{r_b(\tilde{\mathbf{y}}(1))}{\sqrt{\det(\pi D)}} = 1$, we get

$$L_n = n\ell(\tilde{\mathbf{y}}(1)) \left(1 + O\left(\frac{\log^3 n}{\sqrt{n}} \right) \right). \tag{5.29}$$

We terminate the proof of Theorem 5.1 by the fact that $n\ell(\tilde{\mathbf{y}}(1)) = H(X_n)$ (since $\forall c \in \mathscr{A} : u_c(1) = 1$ and $(v_c(1))_{c \in \mathscr{A}}$ is the Markov stationary distribution).

Proof (Proof of Theorem 5.2) The proof of Theorem 5.2 proceeds equivalently except that $E_n = \sum_{b \in \mathscr{A}} E_n^b$ with

$$E_n^b = \sum_{\mathbf{k} + \delta_{ba} \in \mathscr{F}_n} N_{\mathbf{k}}^b \mathbf{P}^{\mathbf{k}} E_{\mathbf{k}}^b. \tag{5.30}$$

The main factor $\mathbf{P}^{\mathbf{k}} B_{\mathbf{k}}^2$ leads to a factor $\exp\left(2nL\left(\mathbf{y}, \frac{1}{2}\right)\right)$. Consideration on the order of the n factors leads to the estimate in $An^{-1}\lambda^{2n}(\frac{1}{2})$.

More extensive studies can be found in our later works in [37–39, 47].

5.4.4 Experimental Results

Figure 5.10 shows $I(X_n, Y_n)$, the discrepancy between L_n and $H(X_n)$, more precisely the mean value of $\log N_{\mathbf{k}} + \log \mathbf{P}^{\mathbf{k}}$ versus the string length n, when X_n is generated by a Markov process of memory 1 based on the statistics of the syntax graph of the sentence *to be or not to be that is the question* (depicted on Fig. 5.9) [38]. As predicted by the conjecture it is quite sub-linear and seems to be in $\log n$. Each point has been simulated 100 times.

5.5 Chapter Summary

In this chapter we presented our work in three additional research subjects, (a) classification encryption via compressed permuted measurement matrices, (b) dynamic classification completeness based on Matrix Completion and (c) encryption based on the Eulerian circuits of original texts.

In the first additional research subject we study the encryption property of Compressive Sensing in order to secure the classification process in Twitter without an extra cryptographic layer. First we performed topic detection based on Joint Complexity and then we employed the theory of Compressive Sensing to classify the tweets by taking advantage of the spatial nature of the problem. The measurements

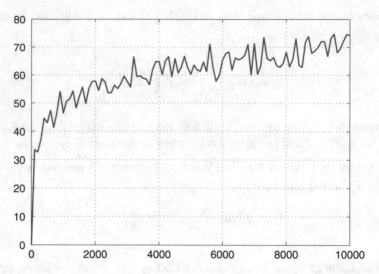

Fig. 5.10 Mutual information $I(X_n, Y_n)$ versus n for the Markov process of Fig. 5.9. Y axis is the mutual information (in bit), X axis is the length of the string up to 10,000

obtained are considered to be weakly encrypted due to their acquisition process, which was verified by the experimental results.

In the second additional research subject we applied a novel framework based on dynamic Matrix Completion to reduce the exhaustive computation of the score matrices during the process of topic detection based on Joint Complexity. The experimental evaluation revealed great performance with low computational complexity.

In the third additional research subject, we evaluated the number of Eulerian circuits that could be obtained by an arbitrary rotation in a Markovian string given to a Markovian type. We also studied the number of Eulerian components that result from random rotations in Markovian strings. The asymptotic behaviour of those quantities was considered, when the size n of a string X tends to infinity. The study of the performance on Markov models with different experiments on real text showed very promising results.

Chapter 6
Conclusions and Perspectives

This book introduced and compared two novel topic detection and classification methods based on Joint Complexity and Compressive Sensing. In the first case, the joint sequence complexity and its application was studied, towards finding similarities between sequences up to the discrimination of sources. We exploited datasets from different natural languages using both short and long sequences. We provided models and notations, presented the theoretical analysis, and we applied our methodology to real messages from Twitter, where we evaluated our proposed methodology on topic detection, classification and trend sensing, and we performed automated online sequence analysis.

In the second case, the classification problem was reduced in a sparse reconstruction problem in the framework of Compressive Sensing. The dimensionality of the original measurements was reduced significantly via random linear projections on a suitable measurement basis, while maintaining an increased classification accuracy. By taking advantage of the weak capability properties we were able to design a secure system for classification without the need of any cryptographic layer.

The empirical experimental evaluation revealed that the methods outperform previous approaches based on bag-of-words and semantic analysis, while the Compressive Sensing based approach achieved a superior performance when compared to state-of-the-art techniques. We performed an evaluation of various datasets in Twitter and compete in the Data Challenge of the World Wide Web Conference 2014, which verified the superiority of our method.

Motivated by the philosophy of posting tweets, a hybrid tracking system is presented, which exploits the efficiency of a Kalman filter in conjunction with the power of Compressive Sensing to track a Twitter user, based on his tweets. The experimental evaluation reveals an increased classification performance, while maintaining a low computational complexity.

© Springer International Publishing AG 2018

D. Milioris, *Topic Detection and Classification in Social Networks*,

DOI 10.1007/978-3-319-66414-9_6

By taking advantage of the spatial nature of the problem, the CS measurements obtained are considered to be weakly encrypted due to their acquisition process. This motivated us to design a secure mechanism for the classification process without an extra cryptographic layer.

In the second additional research subject we applied a novel framework based on dynamic Matrix Completion to reduce the exhaustive computation of the score matrices during the process of topic detection based on Joint Complexity. The experimental evaluation revealed great performance with low computational complexity.

In the third additional research subject, we evaluated the number of Eulerian circuits that could be obtained by an arbitrary rotation in a Markovian string given to a Markovian type. We also studied the number of Eulerian components that result from random rotations in Markovian strings. The asymptotic behaviour of those quantities was considered, when the size n of a string X tends to infinity.

There is a growing interest in statistical methods that exploit various spatio-temporal statistical properties of the datasets to form robust maps of the space. In general, a user's tweets are described by very transient phenomena (frequency, language, list of followers) and is highly time-varying. At the same time, the collection and analysis of data is subject to the quality and relevance of real time search, the correlation information between groups of users, and the analysis of the relationship between members of a group or a community. Thus, the general problem of building a theoretical framework to analyse these data taking into consideration the above limitations opens up exciting research opportunities.

Regarding the theoretical analysis of the Joint Complexity method, we gave a conjecture inspired from the general results about the asymptotic digital trees and suffix tree parameters distribution. We omitted the proof of the conjecture on its variance, which will be studied in a future work.

Regarding the part of Compressive Sensing, in the present work, the unknown class was estimated by performing separate reconstruction for each representative tweet. A straightforward extension would be the use of the joint sparsity structure of the indicator vector s among the reference points for the simultaneous classification.

Besides, a more thorough study should be carried out for the robustness of the inherent encryption property in terms of the several network parameters. The choice of appropriate sparsifying and measurement bases is crucial for an increased classification accuracy. The design of new transform and measurement bases, Ψ, Φ, respectively, being adaptive to the specific characteristics of the data is a significant open problem.

Another interesting use of Joint Complexity method would be the design of a source finding model, based on topic generation processes and the statistics of the obsolescence rates of the topics. We may assume that topic sources are generated on each Twitter user like an i.i.d Poisson process of a specific rate per timeslot studied. Then we have to obtain an obsolescence rate, which defines the number of tweets posted in a timeslot based on the difference between a tweet's time stamp and the time stamp of its source. The source finding model will be also explored in a future work.

Appendix A
Suffix Trees

A.1 Suffix Tree Construction

A suffix tree is composed of nodes, which can have three types:

- internal to the tree, in which case they contain only a list of children (or equivalently outgoing edges), leaves, in which case they contain a pointer to their beginning position in the original string (to save memory space as opposed to storing the whole substring for each leaf). In order to reconstruct the suffix, we simply build a string from the noted position to the end of the original string.
- a special kind of leaf, noted epsilon, which is an empty leaf denoting the end of the string.
- the list of children can be represented as a Map<Character, Node>, where the Character key is the label of the outgoing edge and the value Node is the other extremity of the outgoing edge.

The construction of a Suffix Tree follows:

- a root node is created. We start from the end of the string and add suffixes from each position until we reach the beginning of the string after which time the Suffix Tree construction is complete.
- adding a suffix takes two parameters, the position of the suffix and the character leading to this position, and may have two outcomes:

  - if there was no previous edge with the given label, we simply create the edge and make it point to the given position.
  - if there was already an edge with the same label, then we need to start walking down this edge (and moving along also in the suffix we are trying to add) until we reach a position where the child we are trying to add is not already present,

© Springer International Publishing AG 2018
D. Milioris, *Topic Detection and Classification in Social Networks*,
DOI 10.1007/978-3-319-66414-9

in which case we simply add the edge at the correct sublevel. If at any point during this process we reach a leaf node, we need to expand it and create the corresponding internal nodes as long as the two substrings coincide and sprout new leaves as soon as they differ.

A.2   Suffix Trees Superposition

In order to compute the Joint Complexity of two sequences, we simply need their two respective Suffix Trees.

Comparing Suffix Trees can be viewed as a recursive process which starts at the root of the trees and walks along both trees simultaneously. When comparing subparts of the trees we can face three situations:

- both parts of the trees we are comparing are leaves. A leaf is basically a string representing the suffix. Comparing the common factors of two strings can be done easily by just incrementing a counter each time the characters of both strings are equal and stop counting as soon as they differ. For example, comparing the two suffixes "nalytics" and "nanas" would give a result of 2 as they only share the first two characters; while comparing suffixes "nalytics" and "ananas" would return 0 as they do not start with the same character.
- one subpart is a leaf while the other subpart is an internal node (i.e. a branch). Comparing a non-leaf node and a leaf is done by walking through the substring of the leaf and incrementing the score as long as there is an edge whose label corresponds to the current character of the leaf. Note that the keys (edge's labels) are sorted so that we can stop looking for edges as soon as an edge is sorted after the current character (allowing average sublinear computation). When we reach a leaf on the subtree side, we just need to compare the two leaves (starting from where we are at in the substring and the leaf that we just reached).
- both parts of the trees we are comparing are internal nodes (i.e. we are trying to compare two branches). Comparing two non-leaf nodes is done by initializing a current counter to zero and walking through both subtrees while doing the following at each step:

  - first check whether we have reached the end (epsilon) of one of the subtrees, in which case we can stop comparing and return the current counter.
  - then check whether we have reached two leaves at the same time, in which case we add the current counter to the output of the previously described method for comparing two leaves.
  - check whether we have reached a leaf on one side only, in which case we add the current counter to the output of the previously described method for comparing a leaf with an internal node.
  - keep walking through subtrees as long as we find identical edges (and call ourselves recursively), each time incrementing our internal counter.

References

1. Aardenne-Ehrenfest TV, de Bruijn NG (1951) Circuits and trees in oriented linear graphs. Simon Stevin 28:203–217
2. Aiello LM, Petkos G, Martin C, Corney D, Papadopoulos S, Skraba R, Goker A, Kompatsiaris Y, Jaimes A (2013) Sensing trending topics in twitter. IEEE Trans Multimedia 15(6):1268–1282
3. Allan J (2002) Topic detection and tracking: event-based information organization. Kluwer Academic Publishers, Norwell
4. Baraniuk R (2007) Compressive sensing. IEEE Signal Process Mag 24:118–121
5. Becher V, Heiber PA (2011) A better complexity of finite sequences, abstracts of the 8th. In: International conference on computability, complexity, and randomness, p 7
6. Becker H, Naaman M, Gravano L (2011) Beyond trending topics: real-world event identification on twitter. In: 5th international AAAI conference on web and social media
7. Blei DM, Lafferty JD (2006) Dynamic topic models. In: 23rd ACM international conference on machine learning, New York, pp 113–120
8. Blei DM, Ng AY, Jordan MI (2003) Latent Dirichlet allocation. J Mach Learn Res 3:993–1022
9. Burnside G, Milioris D, Jacquet P (2014) One day in twitter: topic detection via joint complexity. In: Proceedings of SNOW 2014 data challenge (WWW'14), Seoul
10. Cai J, Candès E, Shen Z (2010) A singular value thresholding algorithm for matrix completion. SIAM J Optim 20:1956–1982
11. Candès E, Recht B (2009) Exact matrix completion via convex optimization. Found Comput Math 9:717–772
12. Candés E, Romberg J, Tao T (2006) Robust uncertainty principles: exact signal reconstruction from highly incomplete frequency information. IEEE Trans Inform Theory 52:489–509
13. Cataldi M, Caro LD, Schifanella C (2010) Emerging topic detection on twitter based on temporal and social terms evaluation. In: 10th international workshop on multimedia data mining, New York, pp 1–10
14. Chen S, Donoho D, Saunders M (1999) Atomic decomposition by basis pursuit. SIAM J Sci Comput 20(1):33–61
15. Conti M, Delmastro F, Passarella A (2009) Social-aware content sharing in opportunistic networks. In: 6th annual IEEE communications society conference on sensor, mesh and ad hoc communications and networks workshops, pp 1–3
16. Daemen J, Rijmen V (2002) The design of Rijndael: AES - the advanced encryption standard. Springer, Berlin

17. Diplaris S, Petkos G, Papadopoulos S, Kompatsiaris Y, Sarris N, Martin C, Goker A, Corney D, Geurts J, Liu Y, Point JC (2012) SocialSensor: surfacing real-time trends and insights from multiple social networks. In: NEM summit, pp 47–52

18. Donoho D (2006) Compressive sensing. IEEE Trans Inform Theory 52(4):1289–1306

19. Donoho D, Tsaig Y, Drori I, Starck J (2006) Sparse solution of underdetermined linear equations by stagewise orthogonal matching pursuit. Technical report, Stanford University, CA

20. Fayolle J, Ward MD (2005) Analysis of the average depth in a suffix tree under a Markov model. In: International conference on the analysis of algorithms (AofA), Barcelona, pp 95–104

21. Feng C, Valaee S, Tan Z (2009) Multiple target localization using compressive sensing. In: Proceedings of IEEE GLOBECOM'09, Hawaii

22. Finkel JR, Grenager T, Manning C (2005) Incorporating non-local information into information extraction systems by Gibbs sampling. In: Annual meeting on association for computational linguistics, vol 43, pp 363–370

23. Flajolet P, Sedgewick R (2008) Analytic combinatorics. Cambridge University Press, Cambridge

24. Flajolet P, Gourdon X, Dumas P (1995) Mellin transforms and asymptotics: harmonic sums. Theor Comput Sci 144(1–2):3–58

25. Fung GPC, Yu JX, Yu PS, Lu H (2005) Parameter free bursty events detection in text streams. In: 31st International conference on very large data bases. VLDB Endowment, pp 181–192

26. Garcia-Ruiz MA, Martin MV, Ibrahim A, Edwards A, Aquino-Santos R (2009) Combating child exploitation in second life. In: IEEE international conference on science and technology for humanity, pp 761–766

27. Goethals B (2005) Frequent set mining. Springer, Heidelberg, pp 377–397

28. Grant M, Boyd S (2011) Cvx: Matlab software for disciplined convex programming. Technical report

29. Grewal M, Andrews A (2001) Kalman filtering: theory and practice using MATLAB. Wiley

30. Gyorodi C, Gyorodi R (2004) A comparative study of association rules mining algorithms. In: 1st Joint symposium on applied computational intelligence (SACI 2004)

31. He Q, Chang K, Lim PE (2007) Analyzing feature trajectories for event detection. In: 30th annual international ACM conference on research and development in information retrieval, New York, pp 207–214

32. Hegde N, Massoulié L, Viennot L (2013) Self-organizing flows in social networks. Structural information and communication complexity. Lecture notes in computer science, vol 8179. Springer, New York, pp 116–128

33. Horn RA, Johnson CR (1985) Matrix analysis. Cambridge University Press, Cambridge

34. Ilie L, Yu S, Zhang K (2002) Repetition complexity of words. In: Proceedings of COCOON, pp 320–329

35. Isaev MI (2009) Asymptotic number of Eulerian circuits in complete bipartite graphs. In: 52nd MFTI conference, Moscow

36. Jacquet P (2007) Common words between two random strings. In: IEEE international symposium on information theory, pp 1495–1499

37. Jacquet P, Milioris D (2014) Honeypot design via random Eulerian paths. European Patent No. 14306779.1

38. Jacquet P, Milioris D (2014) Information in strings: enumeration of Eulerian paths and Eulerian components in Markov sequences. In: International conference on probabilistic, combinatorial and asymptotic methods for the analysis of algorithms (AofA'14), Paris

39. Jacquet P, Milioris D (2014) Undetectable encryption based on m-grams permutations. European Patent No. 14305064.9

40. Jacquet P, Szpankowski W (1994) Autocorrelation on words and its applications. Analysis of suffix trees by string-ruler approach. J Combin Theory Ser A 66:237–269

41. Jacquet P, Szpankowski W (1998) Analytical depoissonization and its applications. Theor Comput Sci 201:1–62

42. Jacquet P, Szpankowski W (2004) Markov types and minimax redundancy for Markov sources. IEEE Trans Inform Theory 50(7):1393–1402
43. Jacquet P, Szpankowski W (2012) Joint string complexity for Markov sources. In: 23rd international meeting on probabilistic, combinatorial and asymptotic methods for the analysis of algorithms, vol 12, pp 303–322
44. Jacquet P, Szpankowski W (2015) Analytic pattern matching: from DNA to twitter. Cambridge University Press, Cambridge
45. Jacquet P, Szpankowski W, Tang J (2001) Average profile of the Lempel-Ziv parsing scheme for a Markovian source. Algorithmica 31(3):318–360
46. Jacquet P, Milioris D, Szpankowski W (2013) Classification of Markov sources through joint string complexity: theory and experiments. In: IEEE international symposium on information theory (ISIT), Istanbul
47. Jacquet P, Milioris D, Burnside G (2014) Textual steganography - undetectable encryption based on n-gram rotations. European Patent No. 14306482.2
48. Janson S, Lonardi S, Szpankowski W (2004) On average sequence complexity. Theor Comput Sci 326:213–227
49. Ji S, Xue Y, Carin L (2008) Bayesian compressive sensing. IEEE Trans Signal Process 56(6):2346–2356
50. Jiao J, Yan J, Zhao H, Fan W (2009) Expertrank: an expert user ranking algorithm in online communities. In: International conference on new trends in information and service science, pp 674–679
51. Kondor D, Milioris D (2016) Unsupervised classification in twitter based on joint complexity. In: International conference on computational social science (ICCSS'16), Chicago, IL
52. Lampos V, Cristianini N (2010) Tracking the flu pandemic by monitoring the social web. In: International workshop on cognitive information processing (CIP), pp 411–416
53. Lehmann J, Goncalves B, Ramasco JJ, Cattuto C (2012) Dynamical classes of collective attention in twitter. In: 21st ACM international conference on world wide web (WWW), New York, pp 251–260
54. Leskovec J, Backstrom L, Kleinberg J (2009) Meme-tracking and the dynamics of the news cycle. In: 15th ACM international conference on knowledge discovery and data mining (KDD), New York, pp 497–506
55. Li M, Vitanyi P (1993) Introduction to Kolmogorov Complexity and its Applications. Springer, Berlin
56. Li H, Wang Y, Zhang D, Zhang M, Chang EY (2008) Pfp: parallel fp-growth for query recommendation. In: ACM conference on recommender systems, New York, pp 107–114
57. Li G, Li H, Ming Z, Hong R, Tang S, Chua T (2010) Question answering over community contributed web video. Trans Multimedia 99, 17(4):46–57
58. Markines B, Cattuto C, Menczer F (2009) Social spam detection. In: Proceedings of the 5th ACM international workshop on adversarial information retrieval on the web, New York, pp 41–48
59. Mathioudakis M, Koudas N (2010) Twittermonitor: trend detection over the twitter stream. In International conference on management of data (SIGMOD), New York, pp 1155–1158
60. McKay B, Robinson RW (1995) Asymptotic enumeration of Eulerian circuits in the complete graph. Combinatorica 4(10):367–377
61. Milioris D (2014) Compressed sensing classification in online social networks. Technical report, Columbia University, New York
62. Milioris D (2014) Text classification based on joint complexity and compressed sensing. United States Patent No. 14/540770
63. Milioris D (2016) Classification encryption via compressed permuted measurement matrices. In: IEEE international workshop on security and privacy in big data (BigSecurity'16), INFOCOM'16, San Francisco, CA
64. Milioris D (2016) Towards dynamic classification completeness in twitter. In: IEEE European signal processing conference (EUSIPCO'16), Budapest

65. Milioris D, Jacquet P (2013) Method and device for classifying a message. European Patent No. 13306222.4
66. Milioris D, Jacquet P (2014) Joint sequence complexity analysis: application to social networks information flow. Bell Labs Tech J 18(4):75–88
67. Milioris D, Jacquet P (2014) Secloc: encryption system based on compressive sensing measurements for location estimation. In: IEEE international conference on computer communications (INFOCOM'14), Toronto
68. Milioris D, Jacquet P (2015) Classification in twitter via compressive sensing. In: IEEE international conference on computer communications (INFOCOM'15)
69. Milioris D, Jacquet P (2015) Topic detection and compressed classification in twitter. In: IEEE European signal processing conference (EUSIPCO'15), Nice
70. Milioris D, Kondor D (2016) Topic detection completeness in twitter: Is it possible? In: International conference on computational social science (ICCSS'16), Chicago
71. Milioris D, Tzagkarakis G, Papakonstantinou A, Papadopouli M, Tsakalides P (2014) Low-dimensional signal-strength fingerprint-based positioning in wireless lans. Ad Hoc Netw J Elsevier 12:100–114
72. Murtagh F (1983) A survey of recent advances in hierarchical clustering algorithms. Comput J 26(4):354–359
73. National Bureau of Standards, U. D. o. C., editor. Data encryption standard. Washington DC, Jan 1977
74. Neininger R, Rüschendorf L (2004) A general limit theorem for recursive algorithms and combinatorial structures. Ann Appl Probab 14(1):378–418
75. Niederreiter H (1999) Some computable complexity measures for binary sequences. In: Ding C, Hellseth T, Niederreiter H (eds) Sequences and their applications. Springer, Berlin, pp 67–78
76. Nikitaki S, Tsagkatakis G, Tsakalides P (2012) Efficient training for fingerprint based positioning using matrix completion. In: 20th European signal processing conference (EUSIPCO), Boucharest, Romania, pp 27–310
77. Nilsson S, Tikkanen M (1998) Implementing a dynamic compressed Trie. In: Proceedings of 2nd workshop on algorithm engineering
78. O'Connor B, Krieger M, Ahn D (2010) Tweetmotif: exploratory search and topic summarization for twitter. In: Cohen WW, Gosling S, Cohen WW, Gosling S (eds) 4th international AAAI conference on web and social media. The AAAI Press, Menlo Park
79. Paar C, Pelzl J (2009) Understanding cryptography, a textbook for students and practitioners. Springer, Berlin
80. Papadopoulos S, Kompatsiaris Y, Vakali A (2010) A graph-based clustering scheme for identifying related tags in folksonomies. In: 12th international conference on data warehousing and knowledge discovery, pp 65–76
81. Papadopoulos S, Corney D, Aiello L (2014) Snow 2014 data challenge: assessing the performance of news topic detection methods in social media. In: Proceedings of the SNOW 2014 data challenge
82. Petrovic S, Osborne M, Lavrenko V (2010) Streaming first story detection with application to twitter. In: Annual conference of the North American chapter of the association for computational linguistics, pp 181–189
83. Phuvipadawat S, Murata T (2010) Breaking news detection and tracking in twitter. In: IEEE/WIC/ACM international conference on web intelligence and intelligent agent technology, pp 120–123
84. Porter MF (1997) An algorithm for suffix stripping. In: Readings in information retrieval. Morgan Kaufmann Publishers Inc., San Francisco, pp 313–316
85. Prakash BA, Seshadri M, Sridharan A, Machiraju S, Faloutsos C (2009) Eigenspokes: surprising patterns and scalable community chipping in large graphs. In: IEEE international conference on data mining workshops (ICDMW), pp 290–295
86. Rodrigues EM, Milic-Frayling N, Fortuna B (2008) Social tagging behaviour in community-driven question answering. In: IEEE/WIC/ACM international conference on web intelligence and intelligent agent technology, pp 112–119

87. Sakaki T, Okazaki M, Matsuo Y (2010) Earthquake shakes twitter users: real-time event detection by social sensors. In: Proceedings of the 19th ACM international conference on world wide web (WWW'10), New York

88. Salton G, McGill MJ (1986) Introduction to modern information retrieval. McGraw-Hill, New York

89. Sankaranarayanan J, Samet H, Teitler BE, Lieberman MD, Sperling J (2009) Twitterstand: news in tweets. In: 17th ACM international conference on advances in geographic information systems, New York, pp 42–51

90. Sayyadi H, Hurst M, Maykov A (2009) Event detection and tracking in social streams. In: Adar E, Hurst M, Finin T, Glance NS, Nicolov N, Tseng BL (eds) 3rd international AAAI conference on web and social media. The AAAI Press, Menlo Park

91. Shamma DA, Kennedy L, Churchill EF (2011) Peaks and persistence: modeling the shape of microblog conversations. In: ACM conference on computer supported cooperative work, New York, pp 355–358

92. Szpankowski W (2001) Analysis of algorithms on sequences. Wiley, New York

93. Tata S, Hankins R, Patel J (2004) Practical suffix tree construction. In: 30th VLDB conference, vol 30

94. Teh YW, Jordan MI, Beal MJ, Blei DM (2006) Hierarchical Dirichlet processes. J Am Stat Assoc 101(476):1566–1581

95. Teh YW, Newman D, Welling M (2007) A collapsed variational Bayesian inference algorithm for latent Dirichlet allocation. Adv Neural Inf Process Syst 19:1353–1360

96. Tibshirani R (1996) Regression shrinkage and selection via the lasso. J R Stat Soc Ser B (Methodol) 58(1):267–288

97. Toninelli A, Pathak A, Issarny V (2011) Yarta: a middleware for managing mobile social ecosystems. In: International conference on grid and pervasive computing (GPC), Oulu, Finland, pp 209–220

98. Tropp J, Gilbert A (2007) Signal recovery from random measurements via orthogonal matching pursuit. IEEE Trans Inform Theory 53:4655–466

99. Tutte WT, Smith CAB (1941) On unicursal paths in a network of degree 4. Am Math Mon 48:233–237

100. Tzagkarakis G, Tsakalides P (2010) Bayesian compressed sensing imaging using a gaussian scale mixture. In: 35th international conference on acoustics, speech, and signal processing (ICASSP'10), Dallas, TX

101. Tzagkarakis G, Milioris, D, Tsakalides P (2010) Multiple-measurement Bayesian compressive sensing using GSM priors for doa estimation. In: 35th IEEE international conference on acoustics, speech and signal processing (ICASSP), Dallas, TX

102. Valdis K (2002) Uncloaking terrorist networks, vol 4. First Monday 7

103. Weng J, Lee B-S (2011) Event detection in twitter. In: 5th international conference on weblogs and social media

104. Wheeler DJ, Needham RM (1994) Tea, a tiny encryption algorithm. In: International workshop on fast software encryption. Lecture notes in computer science, Leuven, Belgium, pp 363–366

105. White T, Chu W, Salehi-Abari A (2010) Media monitoring using social networks. In: IEEE second international conference on social computing, pp 661–668

106. Wiil UK, Gniadek J, Memon N (2010) Measuring link importance in terrorist networks. In: International conference on advances in social networks analysis and mining, pp 9–11

107. Xu X, Yuruk N, Feng Z, Schweiger TAJ (2007) Scan: a structural clustering algorithm for networks. In: 13th ACM international conference on knowledge discovery and data mining (KDD), New York, pp 824–833

108. Yang J, Leskovec J (2011) Patterns of temporal variation in online media. In: 4th ACM international conference on web search and data mining, New York, pp 177–186

109. Ziv J (1988) On classification with empirically observed statistics and universal data compression. IEEE Trans Inform Theory 34:278–286

Index

A
Advanced Encryption Standard, 85
AES, 85
Aggregation, 13
API, 6, 44, 47, 49, 51, 61, 62, 72
ASCII, 38
asymptotics, 36, 56
attacks, 73
autocorrelation, 30

B
Badoo, 2
bag-of-words, 5, 10, 74, 79
Barrabes, 2
Bayesian CS algorithms, 65
BCS, 66, 74, 80
BCS-GSM, 66, 74, 80
BNgram, 6, 9, 18, 19

C
cardinality of a set, 5
central tweets, 59, 71
classification encryption, 7
Collapsed Variational Bayesian, 13
compressed trie, 39
Compressive Sensing, 6, 62, 67, 69, 91, 93
computational complexity, 70, 76
congestion, 5
Content sharing, 4
Crowdsourcing, 4
CS, 57, 63, 70
CT, 59, 71
CVX, 78
cyber surveillance, 5

D
Daily-Motion, 2
Data Encryption Standard, 85
DES, 85
df-idft, 19
dictionaries, 5
dimension media servers, 5
Discrete Fourier Transform, 11
DNA, 23
Doc-p, 6, 19
document–pivot, 6, 9
Document–Pivot Topic Detection, 9
Document-Pivot Topic Detection, 9
Double DePoissonization, 32
DP, 53, 79
DPurl, 53, 65, 74, 79
dynamic Matrix Completion, 76
dynamical system, 6
dynamics, 1, 3
DynMC, 76, 78

E
entertainment, 2
Euclidean norm, 78
Eulerian circuits, 7, 69, 70, 82, 83, 92
Eulerian path, 83

F
F-score, 51, 52, 64
Facebook, 2
feature–pivot, 6, 9
FP, 16
FPC, 78

© Springer International Publishing AG 2018
D. Milioris, *Topic Detection and Classification in Social Networks*,
DOI 10.1007/978-3-319-66414-9

Printed in the United States
By Bookmasters